Disasters, Gender and Access to Healthcare

Disasters, Gender and Access to Healthcare: Women in Coastal Bangladesh emphasizes women's experiences in cyclone disasters being confined with gendered identity and responsibilities in developing socio-economic conditions with minimum healthcare facilities.

The study is situated in the coastal region of Bangladesh, considered as one of the most disaster-prone regions in the world. Bangladesh has been working on disaster management for a long time; however, considering gender perspective, the book reveals gaps in plans and raises serious questions about the successful implementation of healthcare strategies after disasters. The book also describes the pre–during–after disaster periods, showing the full picture of a disaster attack in victims' own words. Case studies of seriously affected victims give the reader an opportunity to understand the situations created for women during a disaster attack in a remote area with poor transport and healthcare facilities.

These unique research findings will contribute to the broader context of gender, disaster and health studies. This book will be helpful for university staff and students of different disciplines, including anthropology, disaster management, gender studies, geography and South Asian regional studies, and be invaluable reading for disaster managers, policy makers, aid workers, development partners, NGOs and governments, especially in disaster-prone countries.

Nahid Rezwana is an Assistant Professor in the Department of Geography and Environment at the University of Dhaka in Bangladesh. She was previously an academic tutor at Durham University, where she completed her PhD thesis 'Disasters and access to health care in the coastal region of Bangladesh'. Rezwana was an Assistant Editor for the *Geography and Environment Journal* in 2007 and was awarded the Dhaka University Scholarship and the Doctoral Scholarship from Christopher Moyes Memorial Foundation. She is also an active member of the Bangladesh Geographical Society (BGS).

Routledge International Studies of Women and Place

Series Editors: Janet Henshall Momsen and Janice Monk
University of California, Davis and University of Arizona, USA

For a full list of titles in this series, please visit www.routledge.com/series/SE0406

Disasters, Gender and Access to Healthcare

Women in Coastal Bangladesh

Nahid Rezwana

Routledge
Taylor & Francis Group

LONDON AND NEW YORK

First published 2018 by Routledge

2 Park Square, Milton Park, Abingdon, Oxfordshire OX14 4RN
52 Vanderbilt Avenue, New York, NY 10017

Routledge is an imprint of the Taylor & Francis Group, an informa business

First issued in paperback 2019

British Library Cataloguing-in-Publication Data
A catalogue record for this book is available from the British Library

Library of Congress Cataloging-in-Publication Data
A catalog record has been requested for this book

ISBN: 978-1-138-57354-3 (hbk)
ISBN: 978-0-367-89160-2 (pbk)

Typeset in Times New Roman
by Apex CoVantage, LLC

Contents

Illustrations

Preface

The first two chapters of the book focus on the background to the research and the plan for investigation, and the next six present the findings of the research. Chapter 1 draws upon relevant theories, suggestions and conceptualizations in the prevailing literature to identify the research gap and the ways in which gender analysis in disaster research can be conceptualized. The chapter presents the background to the selection of research questions, study area and methodology for the present research. The selection of Barguna as a suitable study area for the investigation is explained in Chapter 2. This chapter focuses on the vulnerability of the district, describing several major problems in the health sector, along with the poor socio-economic conditions. The methodology and methods of data collection are elaborated in this chapter. This explains all of the steps of data collection and analysis and the challenges faced in conducting fieldwork in the remote district of Barguna. How these challenges were faced and the consequent limitations of the research are also discussed in this chapter.

Chapter 3 and Chapter 4 look at the findings of the data analysis. These chapters focus on the gendered health impacts of disasters, considering both physical and psychological health problems. Drawing on rich empirical evidence from respondents, the chapters reveal the complexity of health problems created in disasters, describing the contexts, reasons and consequences of unequal vulnerability among the victims and the social factors that influence this. Chapter 5 and Chapter 6 then look at the issue of healthcare access after disasters, one of the major questions of this research. These chapters focus on several factors, especially gender and socio-economic conditions, and how these influence the healthcare access of victims, again drawing on respondent's personal experiences. The poor healthcare access of female victims is explored in these chapters. Chapter 7 focuses on the initiatives taken by disaster managers in the pre-, during- and post-disaster periods. It reveals the gaps in disaster management plans and policies and shows the problems faced by the responders after disasters that must be recognized in order to prepare effective disaster management plans. The recommendations of ordinary respondents are also discussed and evaluated in this chapter. Chapter 8, the concluding chapter, summarizes the findings of the research and concludes with recommendations for improving health and healthcare conditions in disaster-prone regions.

Acknowledgements

I would like to start the acknowledgement by expressing earnest gratitude to my creator, Almighty Allah, the Beneficent, the Merciful. It was impossible for me without His will, blessings and help to live this life and follow my dreams.

Being a citizen of a disaster-prone country, I have grown up seeing the devastations of disasters: loss of life, the pain of the victims and their life-long sufferings. As a researcher, I felt the need to do research on disasters and gender, and it became an aim to take these experiences to the wider communities of the world. I have conducted my PhD research on gendered health impacts of disasters from Durham University, UK, and this book is based on that research. However, it was not easy for me alone to fulfil my aim. There are so many people who inspired and helped me to bring this research to life.

Among them, first and foremost, I want to say thank you to Professor Janet Momsen, who has invited and inspired me to write a book on my PhD research. Again, my supervisors Professor Rachel Pain, Professor Peter Atkins and Professor Sarah Curtis, who I believe were sent from heaven for me. I had a dream of research, but I was not able to make it come true without their help and guidance. They went hundreds of miles beyond supervisions to guide, support and show me the way.

I want to express my deepest gratitude to all the members of the Christopher Moyes Memorial Foundation (CMMF), with a special thanks to Mrs Moyes for awarding me the scholarship to conduct PhD research on disaster victims. I am also grateful to the United Nations Development Programme (UNDP), the Institute of Hazard, Risk and Resilience (IHRR), the Geography Department of Durham University, the Wolfson Research Institute for Health and Wellbeing and the committee of the Norman Richardson Postgraduate Research Fund for granting me the funds to conduct my fieldwork, disseminate my research attending international seminars and receive valuable feedback to enrich the research.

Now I would like to thank, from the bottom of my heart, those inhabitants of Barguna who spent long hours and shared their sensitive, sad and very personal memories of disasters with me to make this research worthwhile. I am overwhelmed by their simplicity, friendliness and quality of believing in someone within a very short time. Without their spontaneous participation, it would have been impossible for me to reach the aim of this research.

From my family, I would like to say thank you to my ma and baba, the precious gift from the Almighty Creator. I am so fortunate to have them in my life. I would like to say special thanks to my best friend and husband, Raihan, for being with me from the very beginning of my graduation life and inspiring me to do something special and worthy. Another person, a little one but the most important, sweetest and funniest person in my life, Elma, my daughter, should get a special thanks and blessings from me. She is a great surprise and special gift for me. I love you, my little mummy.

Abbreviations

BBS	Bangladesh Bureau of Statistics
BDRCS	Bangladesh Red Crescent Society
BRAC	Bangladesh Rural Advancement Committee
BMD	Bangladesh Meteorological Department
CCC	Climate Change Cell
CDMP	Comprehensive Disaster Management Programme
CPP	Cyclone Preparedness Programme
CS	Cyclone Shelter
CSBA	Community Skilled Birth Attendant
DGHS	Directorate General of Health Services
DDM	Department of Disaster Management
DRTMC	Disaster Research Training and Management Centre
DMC	Disaster Management Committee
EPI	Expanded Programme on Immunization
ERC	Emergency Relief Chain
FWC	Family Welfare Centre
GAD	Gender and Development
GDP	Gross Domestic Product
HSC	Higher Secondary Certificate
IFRC	International Federation of Red Cross and Red Crescent Societies
INGO	International Non-Governmental Organization
IPCC	Intergovernmental Panel on Climate Change
MoHFW	Ministry of Health and Family Welfare
MBBS	Bachelor of Medicine and Surgery
MCWCs	Mother and Child Welfare Centres
MT	Metric Ton
NGO	Non-Governmental Organization
NIRAPAD	Network for Information, Response and Preparedness Activities on Disaster
OPD	Outpatient Department
PTG	Post-Traumatic Growth
SAMCO	Sub Assistant Community Medical Officer
SAP	South Asia Partnership
SCG	Shelter Coordination Group

SOD	Standing Orders on Disaster
UDMC	Union Disaster Management Committee
UP	Union Parishad
UzDMC	Upazila Disaster Management Committee
WAD	Women and Development
WCD	Women, Culture and Development
WHO	World Health Organization
WID	Women in Development

Glossary of terms

Mauza   Smallest revenue rural geographic unit having jurisdiction list number.

Union   Smallest administrative rural geographic unit comprising mauzas and villages and having a union parishad institution. Every union parishad has an elected union parishad chairman.

Upazila   Several unions form an upazila. Upazila parishad (Local Government) is one of the three tiers of rural local government. Each Upazila has an elected chairman.

Village   Lowest rural geographic unit, either equivalent to a mauza or part of a mauza.

Paurashava 'Paurashava' is a Bengali term, meaning municipality. A municipality is the urban administrative division and has an elected mayor.

1   Introduction

Why is gender analysis important
in understanding the health
impacts of disasters?

1.1 Context

Natural disasters are one of the major problems of humankind (Stromberg 2007). During the recent decade (from 2000 to 2009), the world has faced an annual average occurrence of 387 disasters, with an annual average of 227.5 million victims, with variations in different years. The number of victims was 198.7 million in 2009, 217.3 million in 2010 (Guha-Sapir et al. 2011) and 96.5 million in 2013 (The Watchers 2015). However, these large numbers do not show the total picture of disaster-stricken regions due to the unequal distribution of disaster impacts among and within regions and societies. In 2013, about 88% of disaster mortality was shared by the countries of low income or lower-middle income, and the Asian continent accounted for 90.1% of global disaster victims (The Watchers 2015).

Within societies, disasters affect poor inhabitants more than the rich (Cannon 2002; Few and Tran 2010; Hannan 2002) and have complex gendered consequences (Enarson 2002; Liang and Cao 2014). Disasters have greater impacts on women, on average killing more women than men (Neumayer and Plümper 2007). Reducing the gendered effects of disasters in the world will reduce the number of disaster victims, but this is not an easy task because of its complexity due to different economic and socio-cultural factors. Gender relations are deeply rooted in societies, and gender differences and discrimination prevail prior to the disasters, making inhabitants, especially women, more vulnerable to health consequences (Liang and Cao 2014). Healthcare systems become disturbed after disasters, making it difficult to provide treatment to injured and affected victims (Paul et al. 2011), especially female victims because of different socio-cultural and economic conditions which are revealed in the increased number of miscarriages, and in maternal and infant mortality during and after disasters (Paul et al. 2011; Neumayer and Plumper 2007). Poor women become the most vulnerable victims of disasters, their healthcare needs are especially crucial when pregnant and as new mothers.

Actually, women face several inequalities and inequity in healthcare access during normal times of the year. Previously, I had the opportunity to conduct research on the reproductive healthcare access of Bangladeshi women and found that poor women do not have fair access to healthcare facilities, even when they live just beside the free services (Rezwana 2002). This research raised further questions and made me curious to know what happens to the health condition of female victims who are living in the remote coastal regions after disasters. Five years after this research, Cyclone Sidr hit Bangladesh in November 2007. On Sidr night, I was staying in Dhaka and was not affected directly in the disaster attack, but I was worried the whole night. The next day, I found how secure I had been in my house when I saw all the images of Cyclone Sidr in the newspapers and on television. All the coastal villages were affected, and a great deal of damage had been caused by the strong winds and the storm surge of Cyclone Sidr. Many women tried to survive and save their children whilst fighting with strong winds and the tidal surge. After the cyclone, they were still struggling to survive under the open sky without any help. They needed food, clothes and treatments for them and their children. Many of them were crying, asking for help, and many were in shock, sitting beside their damaged houses, having lost everything, in some cases their whole family. Those images created a personal empathy for me. As a Bangladeshi woman, I felt I might be one of them. Fortunately, I was living in Dhaka, having received an education, and I was earning and had access to locational facilities; but the women in those coastal areas are mostly poor, help-less and dependent on their family economically and socially. They are mothers, daughters and sisters, just like me. They are extremely vulnerable and easy vic-tims of such disasters. Like other cyclones, Cyclone Sidr killed more women than men. However, no detailed information could be found on other health impacts.

Almost every year, the coastal areas of Bangladesh are hit by several cyclones, and there are many news stories reporting on the human impacts. The numbers of deaths and injuries get publicity, but detailed, in-depth reports are rarely found. Though prevailing disaster management plans in Bangladesh have been quite suc-cessful in reducing the number of deaths compared to the previous decades, the health impacts are still high and gendered. Up to now, more women die than men (the ratio of male to female deaths was 1:5 in Cyclone Sidr 2007, Ahmed 2011), and other health impacts are mostly ignored. However, sufficient data and infor-mation are not available on injuries, the after-effects of injuries and the psycho-logical impacts, especially considering gender. The experiences of female victims and the real reasons for their vulnerability to health impacts of disasters do not get priority, despite the urgency.

This is not only in Bangladesh. Gender mainstreaming in humanitarian relief programmes, disaster management plans and programmes is not common in many countries. Reviewing the prevailing literature, it has been revealed that gender and its relation with socio-cultural factors still have not received proper attention in disaster management plans and programmes and in disaster research. In-depth, qualitative research studies focusing on gendered health impacts of disasters at the individual level are rarely found in developed nations and are very rare in less developed nations (Enarson and Meyreles 2004; Haque et al. 2012a). How-ever, emphasizing gender in disaster research and incorporating a gender view in disaster planning, especially in managing health impacts of disasters, is essential to improve the effectiveness of the response programmes (Morrow and Phillips 1999) and, above all, to improve the health conditions of the victims.

So, in my research, I have aimed to reveal the gendered impacts of cyclone disasters on health and access to healthcare following a qualitative research methodology in the coastal region of Bangladesh, which has been identified as one of the most disaster-prone countries in the world (Matin and Taher 2001). Being grounded in the victims' experiences, this research plans to capture many unspoken facts and reasons behind vulnerability and to focus on the gender relations, differences and discrimination embedded in the society. It also plans to reveal the real health conditions in disasters in a less developed nation being influenced by various inter-related socio-cultural, economic and environmental factors.

The implications of this research are cross-disciplinary. This research will be helpful to the researchers, planners and practitioners in the gender, health and disaster fields who aim to help vulnerable citizens and improve their present conditions. It offers them detailed information on the complexity of social factors, gender and culture, captured from the in-depth investigations to consider the real conditions of disaster victims and vulnerable groups of the society. It focuses on the importance of gender equality and equity in planning health sectors and disaster management, and it opens an important arena for the disaster planners to rethink prevailing plans and policies. This research also provides recommendations on healthcare improvement based on empirical data and reviewing the literature, and it specifies important and essential aspects for developing more effective disaster management plans, considering gaps and shortcomings in prevailing plans and policies. Above all, the present research emphasizes the mitigation of disaster health impacts on victims and the decrease of inequality among them.

Research aim

The research aims to make an original contribution to knowledge and understanding by focusing on gender, culture and the social processes affecting disaster vulnerability in the study area. It will emphasize the gendered impacts of disasters on health and healthcare access, seeking to capture local people's (especially women's) views and experiences more effectively from the disaster-prone coastal region of Bangladesh and making recommendations to lessen the effects of disasters on health, decrease the inequality and inequity among victims and improve prevailing disaster management plans.

1.2 Conceptual development and theoretical perspective

The present section has been structured according to key themes in the current research to conduct a critical analysis of existing conceptual and empirical research. Emphasis has been given to recent disaster trends, their gendered impacts on health and healthcare access, and the place of gender in disaster management and development approaches. Through the discussion, the aim is to reveal the significance of gendered analysis in disaster research and to identify research gaps in the prevailing literature.

Disasters and their uneven spatial and social impacts

In recent years, natural disasters have affected an increasing number of people and their livelihoods throughout the world but disasters are uneven spatially and socially. The occurrence of disasters has 'a relation with geographical location and geological-geomorphological settings' (Alcántara-Ayala 2002, p. 108), but 95% of the total death toll is concentrated in the Global South (Alexander 1993). 'While richer nations do not experience fewer natural disaster events than poorer nations, richer nations do suffer less death from disaster. Economic development provides implicit insurance against nature's shocks' (Kahn 2005, p. 271) along with 'higher educational attainment, greater openness, a strong financial sector and smaller government' (Toya and Skidmore 2007, p. 20). Analyzing the past records of natural disasters registered from 1900 until 1999 by regions of the world, it is revealed that Asia faced the highest percentage (42%), followed by the Americas 27%, Europe 13%, Africa 10% and Oceania 8% (Alcántara-Ayala 2002). During the decade 2000–2009 the proportions were Asia 40%, Americas 24%, Africa 17%, Europe 15% and Oceania 4% of disasters (modified from Guha-Sapir et al. 2011). However, comparison according to number of victims due to these disasters shows a sharp difference between the continents. Asia had about 204.29 million victims of 156 disasters during the years 2000–2009, whereas the Americas had only 7.09 million victims in 92 disasters (Guha-Sapir et al. 2011). Even in the recent year 2010, although the Americas were dangerously affected by natural disasters, and this included fatalities, the number of disaster victims remained far higher in Asia (Guha-Sapir et al. 2011). Behind these differences, the reasons include the susceptibility of these continents because of their geographical and geomorphological settings and issues related to the role that social, economic, political and cultural aspects play as factors of vulnerability (Alcántara-Ayala 2002; Few and Tran 2010). According to Horwich (2000), the level of wealth of any economy works as a critical underlying factor in response to natural disasters. Political factors have also been mentioned by other researchers (Albala-Bertrand 1993 cited in Toya and Skidmore 2007, Stover and Vinck 2008). Analyzing the recent database, Kahn (2005) and Toya and Skidmore (2007) show that high income, greater democracy in governance and underlying social/economic fabric increase the safety for all of society. Educational and financial development and size of the government, correlated with income, play a role in determining safety from natural disasters. For example, during Cyclone Larry in Australia, key success factors were the preparedness of the Government and their unity with private sector. Oloruntoba mentioned this success as the result of an effective response programme in northern Australia (Oloruntoba 2010).

The health impacts of disasters: a gendered analysis

The health consequences of disasters cannot be understated – they are massive and long-lasting all over the world (Lai et al. 2003; Keim 2006). Disasters severely affect human health, causing death, injuries and psychological impacts, as well as reducing the provision of healthcare for injuries, diseases and chronic illness due to affected medical infrastructures and facilities (Keim 2006; Schmidlin 2011; Goldman et al. 2014).

The health impacts of disasters

With regard to cyclones, the impacts of disasters start from the pre-disaster phase and continue through to the post-disaster period. Falls, scratches and puncture wounds are created while making preparations, returning from safe shelters or cleaning up after disasters, and strong winds and storm water lead to the most injuries and deaths during the cyclone attack (Goldman et al. 2014). To mitigate these injuries and avoid further deaths, victims should be found and rescued in a timely manner and provided with early care (Johnson and Galea 2009; Fuse and Yokota 2012), but experience from previous disasters show that, despite disaster preparedness, mighty wind storms have significant health impacts. Lai et al. (2003) mentioned in their research that 'Typhoon Nari in Taipei proved that significant damage from natural disasters also can happen to modern healthcare systems in urban areas' (Lai et al. 2003, p. 1109). Again, Hurricane Katrina left considerable health impacts on victims. Sastry and Gregory (2013) revealed 'a significant decline in health for the adult population of New Orleans in the year after the hurricane, with the disability rate rising from 20.6% to 24.6%' (Sastry and Gregory 2013, p. 121). These health impacts become worse if the affected areas lack proper medical response plans, preparations and insufficient local healthcare facilities (Few and Tran 2010; Djalali et al. 2011). Besides the physical injuries, disasters have psychological impacts on victims who are injured and who have observed the disasters. After reviewing the last 40 years of disaster research, Galea et al. mentioned that 'there is substantial burden of post-traumatic stress disorder among persons who experience a disaster' (Galea et al. 2005, p. 84). Evidence from Hurricane Katrina shows the significant presence of post-traumatic stress disorder symptoms among the evacuees (Coker et al. 2006). But the psychological health impacts of disasters are 'exceedingly difficult to assess' due to the spontaneous and chaotic nature of disasters (Bonanno et al. 2010, p. 1). These assessments are mostly conducted in developed countries and rarely get any attention in the low- and middle-income countries (Marrx et al. 2012; Nahar et al. 2014). The consideration of such mental health problems is also absent in the medical response programmes in these countries, leaving them unrevealed and unattended (Nahar et al. 2014).

Gendered health impacts of disasters

Impacts of natural disasters are never entirely determined by nature (Neumayer and Plümper 2007) but are socially constructed under different geographic, social, cultural, political-economic conditions, and they have complex gendered consequences (Enarson 2002). Much research has been conducted on social vulnerability (Skoufias 2003; Keim 2006; Few and Tran 2010; Karim et al. 2014; Webster 2013), with a focus on the complexity of several social factors increasing people's vulnerability to disasters. Few and Tran (2010) found in their research that

'Income-poverty tended to constrain people's ability to prevent impacts, to seek treatment and to withstand disease. But income-poverty also operated in sometimes subtle, sometimes stark inter-linkage with other dimensions' (Few and Tran 2010, p. 536). Gender and cultural differences in gender identity and behaviour are one of the strongest social factors creating differences among members of society (Weisman 1997; Bari 1998; Nahar et al. 2014), increasing women's vulnerability in disasters (Nasreen 2004; MacDonald 2005; Alam and Collins 2010; Rahman 2013; Alam and Rahman 2014).

'Gender is a widely used and often misunderstood term. It is sometimes mistakenly conflated with sex or used to refer only to women' (Momsen 2010, p. 2). Actually, 'sex refers to one's biological category and gender refers to the socially learned behaviours and expectations associated with each sex' (Watkins and Whaley 2000, p. 44). Everywhere gender is crosscut by differences in class, race, ethnicity, religion, age and sexuality or ability (Momsen 2010 and Watkins and Whaley 2000). Gender 'identities' and 'roles' are not fixed and globally consistent. Gender is a process that is performed over time and space, which 'is being regulated and produces subjectivities that are unstable' (Butler 1990 cited in Sultana 2009, p. 436), and gender relations are organized everywhere by dominant power or roles and influence in everyday life in such a way that they become natural and usual to the participants (Bourdieu 1977; Bolin et al. 1998), which often 'places a variety of expectations and constraints on women' (Ahmed et al. 2000, p. 362).

Gender is central to the concerns of this study, as gender differences exist in the experiences of health problems (Watkins and Whaley 2000, p. 43), health behaviours and health consequences of disasters all over the world. Moreover, natural disasters on average kill more women than men and lower their life expectancy (Neumayer and Plümper 2007). Statistics from disaster areas frequently show the disproportionately greater impacts in different forms on women and girls (Fordham 1998). But not all women are universally or identically impacted by every disaster (Enarson 2002); their vulnerabilities vary according to their biological and physiological characteristics (Neumayer and Plümper 2007), socio-economic position in society (Bari 1998), literacy (Saroar and Routray 2012) and opportunities to access disaster information (Ikeda 1995). Reducing the gendered impacts of disasters in the world would significantly reduce the high disaster mortality rate, but this is not an easy task because of the great complexity of economic and socio-cultural factors operating. Generally, it is realized that women in developing countries suffer higher levels of mortality and morbidity in disasters and emergencies, like in Cyclone Gorky and Cyclone Sidr in Bangladesh (Ahmed 2011), the 2004 tsunami Indonesia, Sri Lanka and India (Macdonald 2005) and the Nepal Earthquake in 2015 (Earthquake Report.com 2015), but 'it is not widely acknowledged that a similar model of disproportionate impact also applies in the developed world' (Fordham 1998, p. 127). Examples could be drawn from past disasters when more women than men died or were affected during and after the disasters in economically developed countries, e.g. in the 1948 and 1966 Russian earthquakes (Enarson and Meyreles 2004), the 1982 flash flood in Nagasaki (Ikeda 1995), the Kobe earthquake in 1995 (Neumayer and Plümper 2007), in

Hurricane Katrina and in the different fluvial flood events since 1998 in England and Wales (Tunstall et al. 2006). Apart from these direct impacts of disasters, the secondary health impacts are sometimes more serious than the primary hazard itself, such as lifelong physical and mental health problems. But there is only scant information (Ikeda 1995; Paul et al. 2011) on these secondary impacts, whereas evidence shows that some flood victims in England and Wales (Tunstall et al. 2006) and in Thailand (Overstreet et al. 2011) suffered long-term mental health effects (Paul et al. 2011), and women were more traumatized among the victims (Overstreet et al. 2011).

Though it is expected that high status among women lowers the differential negative effect of natural disasters on them relative to male life expectancy (Neumayer and Plümper 2007), the above discussion shows that in both developed and developing countries, more women became the victims of disasters. Here it should be mentioned that men are also affected by disasters, and, sometimes, their vulnerability is also created due to social attitudes and gendered responsibility (Doyal 2000; Alam and Collins 2010). However, as this research focuses on 'gender equity' (Sen 2002), women, as a high vulnerable group, receive more attention. The following sections, therefore, place women at the centre of the inquiry to find the complex reasons behind their higher rates of death and injury, their greater vulnerability and their health conditions during and after disasters in both developed and developing countries.

The factors creating gendered vulnerability

Everyone faces risk in the face of disasters, but some people are more vulnerable than others (Alwang et al. 2001). 'Vulnerability' is defined as the 'characteristics of a person or group in terms of their capacity to anticipate, cope with, resist and recover from the impact of a natural disaster' (Blakie et al. 1994, p. 9), resulting from a complex interplay of political, economic, social and ideological practices present at a given place and varies with different hazards (Blaikie et al. 1994 cited in Bolin et al. 1998). Again, vulnerability is dynamic (Alwang et al. 2001) because the vulnerability status of a place and inhabitants are influenced and reshaped by the local coupled human-environment system (Turner et al. 2003). Residents who are more socially vulnerable and living in risky locations might be affected by even a moderate hazard and need longer-term recovery (Cutter et al. 2000). People in older, female or minority groups are thought to be more vulnerable to natural disasters (West and Orr 2007). So, vulnerability to disasters varies greatly from the individual level to society as a whole, the position of vulnerable groups depending on various socio-economic factors (Cannon 1994). Among the factors, '[g]ender is one of the most important factors that affects this differential vulnerability in all phases of a disaster' (Ikeda 1995, p. 173), 'along with age, material welfare, level of education, politics, ethnicity' (Alexander 1993 cited by Ikeda 1995, p. 173; Bolin et al. 1998). In this regard, women (Ikeda 1995) and women-headed families are thought to be in the vulnerable group (Wiest et al. 1994) because many decisions are made at the household level, and if women household heads have low educational attainment and less accessibility to proper information, then their vulnerability increases in disasters. Actually, in some countries, women's vulnerability starts from childhood because of poor health due to lower food consumption, as they eat last and least (Enarson 2000). Added to this, poor access to healthcare services (Cannon 2002) makes them physically weaker when disaster strikes. But these culturally created physical weaknesses are by no means the only factors; 'I found that women had died in greater numbers than men, not just because they were physically weaker, but because of Bangladesh's male-dominated social structure, underpinned by religious traditions, which restricts the mobility of women' (Begum 1993, p. 34). The writer had worked as one of the women relief workers after the cyclone of 1991 in Bangladesh.

To prevent this statement being misinterpreted, it should be explained that 'restrictions on the mobility of women' do not mean women that do not go or are not allowed to go to shelters during disasters. On the contrary, the majority of the evacuees in shelters during cyclones in Bangladesh are women, children and old people (Ikeda 1995). The main gap remains rather in women's dependency on others for disaster warnings or information from outside their house (Saroar and Routray 2012) and for the major decisions regarding evacuation before a disaster (Ikeda 1995; Begum 1993). Being in the house, the lack of information makes women more vulnerable and susceptible to panic during any sudden disaster (Ikeda 1995). Women's dress and lack of self-rescue skills also create constraints for them in taking safe shelter, both factors that helped more men to save their own lives in the December tsunami of 2004 in Sri Lanka, for instance (MacDonald 2005). So, the higher vulnerability of women is socially constructed (Neumayer and Plümper 2007) well before the preparation period of any commencing disaster.

When disaster strikes, pre-existing gendered discriminatory practices become intensified (Neumayer and Plümper 2007). Evidence shows that more female than male famine victims die at a very young age because of discriminatory access to food resources during disasters (Neumayer and Plümper 2007). Sometimes, the husband or the head of the family takes the decision that girls should be allowed to die first during a famine (Becker 1996 cited in Neumayer and Plumper 2007). Again, during the 1991 cyclone in Bangladesh, a father was unable to hold on to both his son and daughter from being swept away by a tidal surge, so he released his daughter because of the common son preference (Haider et al. 1993 cited in Neumayer and Plumper 2007). These examples might not happen repeatedly, but they are mentioned here to emphasize that discrimination between males and females still remains even in the vital moment of disasters, illustrating the everyday practices of gender discrimination (Chen et al. 1981).

Again, gender-related issues arise when victims get shelter in refugee camps and temporary housing. Discrimination against women arises in regard to personal safety, food distribution, healthcare and counselling (Enarson 2002). In overcrowded camps, women and girls, if unaccompanied, may become vulnerable to sexual abuse and harassment if law and order are disrupted in the affected area (Wiest et al. 1994 and Baden et al. 1994), as happened in the post-Katrina riots in New Orleans (Neumayer and Plümper 2007) and in the refugee camps of the 2004 tsunami (MacDonald 2005). In addition, women and girls are also more negatively affected by the insufficient health and hygiene conditions in refugee camps and cyclone shelters (Neumayer and Plümper 2007; CCC 2009) and discriminated against in regard to food relief, as happened to the victims of West Bengal flooding (Sen 1988 cited in Neumayer and Plümper 2007) and among Mozambican refugees in Malawi in late 1990s (Ager et al. 1995 cited in Neumayer and Plümper 2007). Because of these complex gender disparities, the female infant mortality rate was found to be double compared to male infants in Burmese refugee camps in Bangladesh (Neumayer and Plümper 2007).

Poverty, gender and impact of disasters

'Poverty is both a cause of vulnerability, and a consequence of hazard impacts' (Cannon 2002, p. 45). Economic conditions or poverty affect people's ability to protect themselves against disasters in regard to poor housing, vulnerable location of residence, limited access to information, social protection, positions and connections with powerful groups (politically or economically strong) and effective preventative or coping strategies (Hannan 2002). Keim (2006) discussed locational vulnerability and argued that '[p]opulations are at risk of death simply by virtue of their physical proximity to low-lying land situated near the coastline' (Keim 2006, p. 40). Few and Tran (2010) mentioned in their research on Vietnam that 'as with most aspects of hazard impact in developing countries, economic factors were seen to have a fundamental role in health-related vulnerability' (Few and Tran 2010, p. 536). Even in developed countries, the suffering of poor inhabitants in disasters is more than the rich. For example, poverty played a major role in Hurricane Katrina, with mortality highest amongst the indigent (Overstreet et al. 2011). At the individual level, women's vulnerabilities to disaster depend on their socio-economic position in society (Bari 1998; Cannon 2002). Consider that about 70% of the poor in the world are women (CCC 2009). 'Everywhere women work longer hours than men' (Momsen 2010, p. 2), but 'in every country the jobs done predominantly by women are the least well paid and have the lowest statuses' (Momsen 2010, p. 3, Lundberg and Parr 2000). Compared to men, more women spend most of their time at unpaid household work (Lundberg and Parr 2000), which keeps them poor, dependent and vulnerable (Bolin et al. 1998). The same or poorer conditions remain for most female-headed families, as over 95% of them are living below the poverty line (Cannon 2002). Women in these families are mostly divorced or widowed, and they experience higher levels of vulnerability than male-headed families (Enarson and Morrow 1998). Social isolation, lack of assistance during an evacuation, lack of money for emergency support, violence and insecurity make these women vulnerable during disasters (Enarson 2002). Even in male-headed families, 'underlying social structures serve to reinforce divisions before, during and after disaster strikes in developed as well as developing countries' (Fordham 1998, p. 127). Mostly in every society, women play triple roles as caregivers, producers (food production) and community actors (Moser 1993; Enarson 2000; Nelson et al. 2002), 'in addition to an expanding involvement in paid employment' (Momsen 2010, p. 2). These imposed responsibilities for the home and its belongings increase women's vulnerability during disasters in any country of the world (Hyndman 2008; Fordhum 1998; Bari 1998). This can result in broken marriages in the developed world (Fordham 1998), but it can become a life risk for women in developing countries. As a result of personal poverty, the insecurity of losing shelter and honour after a divorce and the fear of being blamed or even punished by family and society (Bari 1998), women in the developing countries take risks during disasters to save even very low priced property like chickens (Akter 1992 cited in Ikeda 1995). 'The everyday lives of poor and low income women reflect in stark relief their increasing economic insecurity, often exacerbated by their sole responsibility for maintaining families' (Enarson and Morrow 1998, p. 160). High levels of pre-disaster poverty (Nelson et al. 2002), their secondary status in the labour force (Momsen 2010), extensive informal-sector work like household works and lack of rights to resources (Sultana 2010), and their extensive domestic responsibilities clearly make women economically vulnerable long before a natural disaster strikes (Enarson 2000).

Again, women's workload increases dramatically after disasters because of increased household works and caregiving responsibilities (Hannan 2002; Nelson et al. 2002). Even young girls are forced to drop out of school to take over some family work (Hannan 2002). Again, women also lose their essential productive resources, like land, gardens, animals and jewellery (Hyndman 2008; Hannan 2002), during disasters, which makes them more economically dependent on family income (Khondoker 1996 cited in Cannon 2002, p. 48). All these changes make women more poor and vulnerable during the post-disaster recovery phase (Wiest et al. 1994).

Women's health and special needs in disasters

Disaster health impacts on women should be considered specially because women's health is affected by gender-specific biological or reproductive factors, and their access to healthcare depends on their social and economic status (Paolisso and Leslie 1995). Generally, women's healthcare needs are categorized as 'reproductive' and then 'all other' (Clancy and Massion 1992), though women's health includes problems beyond reproduction. In some cases, gender differences embedded into the society place emphasis on women's reproductive roles rather than other health problems and lower their self-respect due to the lack of decision-making power related to marriage and childbirth (Ahmed et al. 2000). Gender differences exist significantly in the experience of health problems, and even the 'need for medicines is gendered, as women's experience of illness differs from that of men' (Momsen 2010, p. 79). Malnutrition and chronic energy deficiency due to inadequate food are common among girls and women in the developing countries (Mukuria et al. 2005; Momsen 2010; Baden et al. 1994), and this health status increases the risk of miscarriage and 'makes women more susceptible to diseases like malaria, tuberculosis, diabetes, hepatitis and heart diseases' (Momsen, 2010, p. 81). So, women victims should be considered for their special healthcare needs for childbirth and related health hazards because when the basic healthcare is severely damaged, the number of miscarriages and maternal and infant mortality rate rise (Paul et al. 2011; Neumayer and Plümper 2007). Both in developed or developing countries, disaster-affected areas face disturbed healthcare systems in most cases. Damaged healthcare facilities like infrastructures, emergency services and machines, lack of electricity and water and the absence of physicians all make it difficult to provide the necessary healthcare to injured victims and chronic disease patients (Paul et al. 2011). For example, six months after Hurricane Katrina in 2005, only one-third of the hospitals in the affected parts of New Orleans were functioning and then at only 20% of their former capacity (Overstreet et al. 2011). Inadequate medical services, including financial resources to access medical facilities, compound the problem for women and girls in developing countries (Sultana 2010). When surrounded by polluted floodwater, women become vulnerable to various illnesses like skin problems, typhoid, cholera and reproductive health problems due to their household washing and sanitation (Rashid 2000 cited in Sultana 2010, p. 48). The lack of adequate safe and private sanitation is another gendered crisis during environmental disasters (Momsen 2010; Sultana 2010; Hannan 2002), and, in some societies, these health problems and additional healthcare needs of women are not shared with male doctors and volunteers; rather, they suffer or even die (Begum 1993; Momsen 2010; Baden et al. 1994). In disasters when women's healthcare problems and vulnerability increase by their social status and gender role, extra attention should be given to their special needs. So, health interventions must take into consideration the social, behavioural and environmental contexts that encompass the lives of women and their ability to address these problems (Paolisso and Leslie 1995; Anwar et al. 2011). Puentes-Markides 1992 in her paper discussed accountability at the institutional level. Affordability (characteristics of clients) and acceptability (behaviour of clients) at the individual/family level should also be considered for analysis of the healthcare utilization of women (Puentes-Markides 1992). But even after the recent cyclone disaster in 2007 in Bangladesh 'no comparative health survey was carried out to examine health impacts of this disaster' (Paul et al. 2011, p. 851), showing limited statistics on other health problems (Ikeda 1995; Paul et al. 2011). In regard to mental health problems, emotional and psychological stresses are always higher among women victims (Sultana 2010; Fordhum 1998; Momsen 2010; Bolin et al. 1998) due to an increase of domestic violence and conflict and the loss of shelters and property, which also increases the sense of powerlessness and marginalization (Hossain et al. 1992 cited in Sultana 2010; Wiest et al.1994 and Baden et al. 1994). Women receive less and poorer-quality healthcare in comparison to men, e.g. in India, girls are 40 times less likely than boys to be taken to a hospital (Momsen 2010; Cannon 2002), and, when a natural disaster strikes, these pre-existing discriminatory practices become aggravated, and negative health impacts on women and girls are intensified (Neumayer and Plümper 2007).

Another important issue is the health impacts of climate change on women, which has been suggested for future research by Haque and his colleagues (2012a) and Sultana (2010). The global climate is changing, and both developed and developing countries are already facing the adverse effects, such as prolonged floods and severe drought in South Asia and Africa, heat waves in Europe and devastating hurricanes in America (Haque et al. 2012a). And this climate change is projected to increase threats to human health, particularly in lower-income populations and in tropical/subtropical countries (IPCC 2001 cited in Haque et al. 2012a) compounded by poor socio-economic conditions and weak health systems (Haque et al. 2012a). Climate change with increasing numbers of disasters (Ali 1996; Agrawala et al. 2003; Karim and Mimura 2008 and Sultana 2010) and outbreaks of different diseases (Haque et al. 2012a) will intensify the prevailing health problems of women (Nelson et al. 2002). Nonetheless, little importance has been given regarding the effect of climate change on women's health due to 'gender-blindness' in policy development (Nelson et al. 2002), the subject of the next section.

Prevailing initiatives and the gap between plans and implementations

Natural disasters are a great concern around the world. 'The frequency and scale of humanitarian emergencies resulting from natural disasters has increased' (Oloruntoba 2010, p. 85), and disaster preparedness and prevention are now a priority concern for the affected countries, donor agencies and implementing organizations (Kahn 2005). But disasters like Hurricane Katrina, Hurricane Rita, the Australian bushfires in 2009, Typhoon Nari, the Bam earthquake, the tsunami in 2004, Cyclone Sidr, Cyclone Nargis, the Orissa floods in 2008 and the Japan tsunami 2011 (Lai et al. 2003; MacDonald 2005; Litman 2006; Stover and Vinck 2008; Fisher 2010; Oloruntoba 2010; Djalali et al. 2011; Fuse and Yokota 2012; Phalkey et al. 2012; Government 2008; CDMP 2014) showed that even good disaster preparation can fail and raise questions about the success of the emergency response programmes. According to these researches, depending on the socio-economic and political condition of the country, emergency programmes were more or less influenced by disrupted transport, insufficiency of medical facilities and resource scarcity, coordination difficulties among the responders and an overall absence of disaster management plans and policies focusing on social issues. Social issues like gender are sometimes mentioned in the disaster plans and policies without considering the strong relation of culture and gender, which becomes a constraint for the successful implementation of response plans. Even 'well-intentioned work of development workers and humanitarian actors can unwittingly reproduce and perpetuate existing gender, racial and geographical hierarchies by uncritically promoting certain kinds of projects i.e., sewing for poor, conflict-affected women' (de Alwis and Hyndman 2002, p. 118 cited in Hyndman 2008). So, response programmes should be gender-sensitive to optimize the effectiveness of the programmes, whereas experience of previous disasters shows that 'the gender dimensions of the disaster are little acknowledged or understood' (Pittaway et al. 2007, p. 307), and gender needs are not focused in response programmes (Moser 1993; Enarson 2000; Pittaway et al. 2007), as female victims need clothes and female doctors and medicine (Begum 1993) and diapers for their infants (Zilversmit et al. 2014), which did not get proper priority in relief efforts. Again, 'women relief workers are important to female survivors' (Enarson 2000, p. 28), and female victims feel uneasy discussing their needs and health problems with male relief workers and doctors. Still, the presence of female relief workers and medical staff is insufficient and sometimes missing altogether in the field, whereas the efficiency of relief distribution increased when female workers were included after a major Turkish earthquake and a cyclone in 1991 in Bangladesh (Enarson 2000; Begum 1993).

Again, gender inequities also prevail in distributing relief, as relief does not reach women as quickly as men (Ikeda 1995), and 'women were pushed down and left behind in the rush' in the relief collection queue (Begum 1993, p. 36), although women are sometimes cynically presented to relief centres and relief lines to take advantage of the female victim image and maximize benefits (Enarson and Morrow 1998). Again, distributing the relief, low-income, female-headed families (collective and multi-family households) were found to be disadvantaged by the agencies assuming one head of household at each address (Morrow and Enarson 1994 cited in Enarson and Morrow 1998), and, anyway in the US, women-owned small businesses receive disproportionately low government loans (Nigg and Tierney 1990). Again, after the tsunami in 2004, it became difficult for the women victims in some areas of South Asian countries to receive emergency supplies due to gender-blindness in response programmes, recognizing only men as the household head (Oxfam 2005 cited in MacDonald 2005), which sometimes excluded women who had lost husbands (Dominelli 2013). Fisher mentioned in her research that increase in violence against women after tsunami 2004 in Sri Lanka 'could have been lessened by more gender-sensitive disaster management' (Fisher 2010, p. 913).

Again, after disasters, both men and women suffer from loss of employment, but women are slower to return to paid work (Enarson 2000; Nelson et al. 2002) or restore their livelihood (Baden et al. 1994 cited in Cannon 2002) because most recovery programmes are targeted mainly at men's paid work, e.g. food for work after Hurricane Mitch (Enarson 2000), work opportunities after drought-stricken areas in Morocco (Nelson et al. 2002) and Australian bush fires (Enarson 2000). Dominelli mentioned in her research that women were absent as active participants in all stages of the humanitarian aid giving processes in Sri Lanka after the 2004 tsunami and 'often miss out on aid entitlement unless they find people to advocate on their behalf or form groups to undertake such action themselves. Donors and recipients alike collude in this *gender silence*' (Dominelli 2013, pp. 77–93).

Understanding gender, culture and disasters: the research gap

The above discussion has shown that, though disaster health impacts vary significantly according to the gendered identity of the victims, gender has not yet received proper attention in prevailing disaster management plans, programmes and policies. At this point, the question arises how gender has been considered in disaster research. Actually, gender, as a category in the analysis of pre-during-post-disaster situations has only emerged recently (Bari 1998), and, even today, gender does not get proper attention in disaster research (Enarson 1998). 'Gender has not yet been mainstreamed in humanitarian relief, nor integrated into the research and field projects, nor included in course work . . .' at most educational institutions (Enarson and Meyreles 2004, p. 53), and nor has it received adequate attention in policies and strategies of disaster management in most countries (Hannan 2002). 'Many research for development efforts focusing on climate change issues do not take gender considerations into account' (Chaudhury et al. 2012, p. 2). Gender is ignored in disaster management approaches like the Kyoto Protocol and the UN Framework Conventions on Climate Change (CCC 2009). Again, besides the irregular presence, imbalance in gender priorities and involvement is also revealed in gender and disaster research. 'Historically, the dominant theoretical perspectives, research strategies and guiding questions in disaster social science have been determinedly male-oriented if not male-dominated' (Enarson and Meyreles 2004, p. 49), and women are not involved as equal partners in disaster mitigation, community-based planning and preventative and coping strategies during the post-disaster period (Enarson 2002). Again, insufficient attention to gender identities and relations has also been observed in the disaster research. 'Men' and 'women' are not researched in terms of their own identities and needs. Disaster research generally provides 'only basic information on gender "differences"', and does not 'engage in any thorough explanation or analysis of women's experiences or perspectives in disasters' (Fothergill 1998, p. 12). It ignores women's real needs in the wake of disasters (Begum 1993). How gender affects men's lives in the disaster context is rarely considered as a key research question in highly developed nations and is very rare in the less developed nations (Enarson and Meyreles 2004).

Besides analysis of disaster events at the individual level for victims, research on the experiences and views of responders and planners is rare in the less developed nations (Enarson and Meyreles 2004). Haque et al. 2012a mentioned that most of the studies on climate change impacts were based on secondary data, except for a very few studies (Haque et al. 2012a). In reviewing the literatures, the present study also revealed that the majority of research follows a quantitative research methodology and methods without detailed investigation and analysis at the individual level.

However, it is argued in this research that understanding specific gender dimensions in disasters (Cannon 2002) and the relation of gender with the complex interplay of power, resources, privileges, stratification and social opportunities at the individual level will increase the effectiveness of disaster management plans (Morrow and Phillips 1999). But focusing on gender alone for disaster management is not sufficient, because 'gender relations are an interactive connection and distinction among people (and groups of people) – what happens to one group in this affects the others and is affected by them' (Connell 2003, p. 3). Hyndman in her research argues that we should more strongly associate feminist disaster research with development approaches (Hyndman 2008, p. 101), and Fisher also recommends 'gender mainstreaming' as a strategy to incorporate gender views in disaster management (Fisher 2010). Therefore, with the aim of revealing the development approach for gender-sensitive disaster management planning, prevailing literatures have been reviewed, and a brief discussion is given in the following paragraphs.

Historically, special attention for 'women' in development projects emerged in the early 1970s after Ester Boserup published her book on women and development (Momsen 2010). From then onwards, different approaches have been formulated and practised to establish a focus on women in development policy. 'Gender' and gender relations in development were important by the end of the twentieth century, when previous approaches to development involving a focus on women were combined into gender and development (GAD) (Momsen 2010). GAD examines 'power relations between women and men, shifting the focus away from 'women' alone' (Hyndman 2008, p. 104). Before that, 'women or gender is simply grafted in onto existing planning traditions without any fundamental changes' (Moser 1993, p. 86), but the goals of development planning could not be fulfilled by this 'add-women-and-stir method' (Boxer 1982 cited in Moser 1993, p. 87), and planners recognized the need to develop gender planning as a planning tradition in its own right and both genders, male and female, should be equally focused on in the planning. Gender equality which recognizes that 'men and women often have different needs and priorities, face different constraints and have different aspirations' (Momsen 2010, p. 8) is needed in policy to enhance development. So, the term 'mainstreaming gender' became well known in development approaches after the 1995 UN Fourth World Conference on Women held in Beijing. The mainstreaming gender equality approach 'tries to ensure that women's as well as men's concerns and experiences are integral to every step of planning' and confirm that 'inequality is not perpetuated' (Momsen 2010, p. 15).

These development approaches Women in Development (WID), Women and Development (WAD) and GAD could be very useful to improve health conditions during and after disasters in a disaster-struck region, given the earlier discussion in this chapter. However, these approaches have been criticised for 'not taking culture adequately into account' (Bhavnani et al. 2003, p. 6). The vulnerability of women to disasters is primarily 'cultural and organizational rather than biological or physiological' (Wiest et al. 1994), and 'women's status indicators are likely to influence the health of women by the way they limit their socio-economic opportunities, fertility choices, nutritional status, and access and utilization of health services, along with other factors' (Mukuria et al. 2005). So, culture should be taken into account in development approaches, which was highlighted by Sen in his research: 'the cultural dimension of development requires closer scrutiny in development analysis' (Sen 2004, p. 37). Bhavnani et al. (2003) also emphasized culture in their Women, Culture, Development (WCD) approach 'as lived experience rather than as a static set of relationships permits an opening of new avenues for development' (Bhavnani et al. 2003, p. 6). They mentioned that the 'WCD lens brings women's agency into the foreground (side by side with and within, the cultural, social, political and economic domains) as a means for understanding how inequalities are challenged and reproduced' (Bhavnani et al. 2003, p. 8). According to this intersectional approach, ethnicity, religion, age, class and gender are the aspects of women's lives that cannot be omitted from any analysis or practice.

Referring the above discussion, it could be concluded that one perspective or approach on gender is insufficient to analyze and improve the health conditions in disaster-struck areas. Gaining the advancement of women and achievement of equality to men requires two approaches – integrating gender concerns through-out the planning/policy and focusing on the special concerns of women (March et al. 1999). This research therefore prefers to integrate two approaches for its exploration; it argues that the 'gender mainstreaming' and 'Women, Culture and Development' approaches are most useful for disaster studies and management. It also suggests that these approaches will be helpful to develop gender-sensitive disaster management plans (Fisher 2010), which focuses on gender equality and equity, with special concerns for women's vulnerability considering their age, race, religion and culture and geographical location, to reduce the imbalance in the health impacts of disasters. Here, gender equity is especially emphasized in disaster health planning, as 'it includes concerns about achievement of health and the capability to achieve good health, not just the distribution of health care. But it also includes the fairness of processes and thus must attach importance to non-discrimination in the delivery of healthcare' (Sen 2002, p. 665). Again, to gain gender equality, men's and boys' participation is required, as they are 'una-voidably involved in gender issues' (Connell 2003, p. 3), and their relation with women should be seen as co-operation, supportive to each other for familial and social bonds, rather than only exploitation, subordination and conflict (Bhavnani et al. 2003, p. 6).

So, according to the above argument, more research is needed on gender and disasters which emphasizes both men's and women's experiences and needs. As Fisher states, 'Disasters are therefore inherently social processes and as such they impact upon the individual differently' (Fisher 2010. p. 904). Further, 'women's lives are qualitatively different from men's lives' (Monk and Hanson 2008, p. 35) in relation to culture and society, and in most societies, their status is not equal to men, but rather far behind. So, individuals' voices, and especially those of more vulnerable groups, should be listened to and considered, both in research and in the process of disaster management planning, which may offer 'many possible paths to disaster risk reduction in the globalizing world' (Enarson and Meyreles 2004, p. 51).

1.3 Conclusion

Reviewing prevailing literatures has enabled me to focus my research to build on the understanding that the health impacts of disasters are socially uneven and gen-dered, and disaster vulnerability differs with gender identity and relations along with other social and economic factors. It also strongly suggests the need for qual-itative research at the individual level in the most disaster-prone areas (Enarson and Meyreles 2004), which will capture the everyday realities and vulnerabilities of men and women, their experiences of disasters and their survival and resilience in the face of disaster health impacts.

Being inspired and influenced by the readings and findings of the literature review, I have selected the coastal region of Bangladesh, one of the most disaster-prone regions in the world, to conduct my research with a qualitative research methodology and methods. I go on to elaborate on these in the next chapter.

References

Ager, A., Ager, W. and Long, L. (1995). "The differential experience of Mozambian refugee women and men." *Journal of Refugee Studies* 8(3): 265–287.

Agrawala, S., Ota, T., Ahmed, A. U., Smith, J. and Aalst, M. V. (2003). "Development and climate change in Bangladesh: Focus on coastal flooding and the Sundarbans." OECOD, France.

Ahmed, N. (2011). "Gender and climate change: Myth vs. reality." *End Poverty in South Asia: Promoting Dialogue on Development in South Asia*, The World Bank. http://blogs.worldbank.org/endpovertyinsouthasia/gender-and-climate-change-myth-vs-reality, accessed on 30/5/2015.

Ahmed, S. M., Adams, A. M., Chowdhury, M. and Bhuiya, A. (2000). "Gender, socio-economic development and health-seeking behaviour in Bangladesh." *Social Science & Medicine* 51(3): 361–371.

Akhter, F. (1992). "How Women Cope: Women Are Not Only Victims." In Hossain, C. P. Dodge & F.H. Abed (eds.) *From Crisis to Development: Coping With Disasters in Bangladesh*. Dhaka, University Press Ltd: 59–65.

Alam, E. and Collins, A. E. (2010). "Cyclone disaster vulnerability and response experiences in coastal Bangladesh." *Disasters* 34(4): 931–954.

Alam, K. and Rahman, M. H. (2014). "Women in natural disasters: A case study from southern coastal region of Bangladesh." *International Journal of Disaster Risk Reduction* 8: 68–82.

Albala-Bertrand, J. M. (1993). *Political Economy of Large Natural Disasters*. New York, Oxford University Press Inc.

Alcántara-Ayala, I. (2002). "Geomorphology, natural hazards, vulnerability and prevention of natural disasters in developing countries." *Geomorphology* 47(2): 107–124.

Alexander, D. E. (1993). *Natural Disasters*, Berlin, Springer Science & Business Media.

Ali, A. (1996). "Vulnerability of Bangladesh to climate change and sea level rise through tropical cyclones and storm surges." *Water, Air, & Soil Pollution* 92(1): 171–179.

Alwang, J., Siegel, J. and Jorgensen, S. L. (2001). "Vulnerability: A view from different disciplines." Social protection discussion paper series, The World Bank, 115.

Anwar, J., Mpofu, E., Mathews, L., Shadoul, A. F. and Brack, K. E. (2011). "Reproductive health and access to healthcare facilities: Risk factors for depression and anxiety in women with an earthquake experience." *BMC Public Health* 11(1): 523. www.biomedcentral.com/1471-2458/11/523.

Baden, S., Green, C., Goetz, A. M. and Guhathakurta, M. (1994). "Background report on gender issues in Bangladesh." Institute of Development Studies, University of Sussex. IDS. Report 26.

Bari, F. (1998). *Gender, Disaster, and Empowerment: A Case Study From Pakistan*, USA: Praeger Publishers.

Becker, J. (1996). *Hungry Ghosts: China's Secret Famine*. New York, John Murray.

Begum, R. (1993). "Women in environmental disasters: The 1991 cyclone in Bangladesh." *Gender & Development* 1(1): 34–39.

Bhavnani, K., Foran, J. and Kurian, P. A. (2003). "An Introduction to Women, Culture and Development." In Bhavnani, K., Foran, J. and Kurian, P. A. (eds.) *Feminist Futures: Re-Imagining Women, Culture and Development*. London, Zed Books: 1–21.

Blaikie, P., Cannon, T., Davis, I. and Wisner, B. (1994). *At Risk: Natural Hazards, People's Vulnerability and Disasters*. London, Routledge, Taylor & Francis Group.

Bolin, R., Jackson, M. and Crist, A. (1998). "Gender Inequality, Vulnerability, and Disaster: Issues in Theory and Research." In Enarson, E. P and Morrow, B. H. (eds.) *The Gendered Terrain of Disaster: Through Women's Eyes*, New York, Praeger: 27–44.

Bonanno, G. A., Brewin, C. R., Kaniasty, K. and Greca, A. M. L. (2010). "Weighing the costs of disaster consequences, risks, and resilience in individuals, families, and communities." *Psychological Science in the Public Interest* 11(1): 1–49.

Bourdieu, P. (1977). *Outline of a Theory of Practice*. Cambridge, Cambridge University Press.

Boxer, M. (1982). "For and About Women: The Theory and Practice of Women's Studies in the United States." In Keohane, N., Rosaldo, M. and Gelpi, B. (eds.) *Feminist Theory: A Critique of Ideology*, USA, Harvester Press, 237–271.

Butler, J. (1990). *Gender Trouble: Feminism and the Subversion of Identity*. New York, Routledge.

Cannon, T. (1994). "Vulnerability analysis and the explanation of 'natural' disasters." in A. Varley (eds.) *Disasters, development and the environment*. Chichester, UK, John Wiley & Sons:13–30.

Cannon, T. (2002). "Gender and climate hazards in Bangladesh." *Gender & Development* 10(2): 45–50.

CCC. (2009). *Climate Change, Gender and Vulnerable Groups in Bangladesh*. Dhaka, Climate Chamge Cell, DoE, MoEF, Component 4b, CDMP, MoFDM: 1–82.

CDMP. (2014). *Assessment Stakeholder's Role in Preparation For and Facing the Tropical Storm Mahasen*. Bangladesh, Ministry of Disaster Management and Relief.

Chaudhury, M., Kristjanson, P., Kyagazze, F., Naab, J. and Neelormi, S. (2012). "Participatory gender-sensitive approaches for addressing key climate change-related research issues: Evidence from Bangladesh, Ghana, and Uganda." CCAFS Working Paper 19, CGIAR Research Program on Climate Change, Agriculture and Food Security, Copenhagen.

Chen, L. C., Emdadul, H. and D'Souza, S. (1981). "Sex bias in the family allocation of food and health care in rural Bangladesh." *Population and Development Review* 7(1): 55–70.

Clancy, C. M. and Massion, C. T. (1992). "American women's health care." *JAMA: The Journal of the American Medical Association* 268(14): 1918–1920.

Coker, A. L., Hanks, J. S., Eggleston, K. S., Risser, J., Tee, P. G., Chronister, K. J., Troisi, C. L., Arafat, R. and Franzini, L. (2006). "Social and mental health needs assessment of Katrina Evacuees." *Disaster Management & Response* 4(3): 88–94.

Connell, R. W. (2003). *The Role of Men and Boys in Achieving Gender Equality*, United Nations, Brazil, Division for the Advancement of Women.

Cutter, S. L., Mitchell, J. T. and Scott, M. S. (2000). "Revealing the vulnerability of people and places: A case study of Georgetown County, South Carolina." *Annals of the Association of American Geographers* 90(4): 713–737.

de Alwis, M. and Hyndman, J. (2002). *Capacity-building in Conflict Zones: A Feminist Analysis of Humanitarian Assistance in Sri Lanka*. Colombo, International Centre for Ethnic Studies.

Djalali, A., Khankeh, H., Ohlen, G., Castren, M. and Kurland, L. (2011). "Facilitators and obstacles in pre-hospital medical response to earthquakes: A qualitative study." *Scandinavian Journal of Trauma, Resuscitation and Emergency Medicine* 19(1): 30.

Dominelli, L. (2013). "Gendering Climate Change: Implications for Debates, Policies and Practices." In Alston, M. and Whittenbury, K. (eds.) *Research, Action and Policy: Addressing the Gendered Impacts of Climate Change*. London, Springer: 77–93.

Doyal, L. (2000). "Gender equity in health: Debates and dilemmas." *Social Science & Medicine* 51(6): 931–939.

Earthquake Report.com. (2015). http://earthquake-report.com/2015/04/25/massive-earthquake-nepal-on-april-25-2015, accessed on 6/6/2015.

Enarson, E. P. (1998). "Through women's eyes: A gendered research agenda for disaster social science." *Disasters* 22(2): 157–173.

Enarson, E. P. (2000). *Gender and Natural Disasters*. Geneva, ILO.

Enarson, E. (2002). "Gender Issues in Natural Disasters: Talking Points on Research Needs." *Crisis, Women and Other Gender Concerns*. Working paper, ILO, Geneva: 5–12.

Enarson, E. P. and Meyreles, L. (2004). "International perspectives on gender and disaster: Differences and possibilities." *International Journal of Sociology and Social Policy* 24(10/11): 49–93.

Enarson, E. P. and Morrow, B. H. (eds.) (1998). *The Gendered Terrain of Disaster*. New York, Praeger.

Few, R. and Tran, P. G. (2010). "Climatic hazards, health risk and response in Vietnam: Case studies on social dimensions of vulnerability." *Global Environmental Change* 20(3): 529–538.

Fisher, S. (2010). "Violence against women and natural disasters: Findings from post-tsunami Sri Lanka." *Violence Against Women* 16(8): 902–918.

Fordham, M. H. (1998). "Making women visible in disasters: Problematising the private domain." *Disasters* 22(2): 126–143.

Fothergill, A. (1998). "The Neglect of Gender in Disaster Work: An Overview of the Literature." In Enarson, E. P. and Morrow, B. H. (eds.) *The Gendered Terrain of Disaster: Through Women's Eyes*, New York, Praeger: 11–25.

Fuse, A. and Yokota, H. (2012). "Lessons learned from the Japan earthquake and tsunami, 2011." *Journal of Nippon Medical School* 79(4): 312–315.

Galea, S., Nandi, A. and Vlahov, D. (2005). "The epidemiology of post-traumatic stress disorder after disasters." *Epidemiologic Reviews* 27(1): 78–91.

Goldman, A., Eggen, B. and Murray, V. (2014). "The health impacts of windstorms: A systematic literature review." *Public Health* 128(1): 3–28.

Government. (2008). *Cyclone Sidr in Bangladesh: Damage, Loss and Needs Assessment For Disaster Recovery and Reconstructions*. Bangladesh, Government of Bangladesh.

Guha-Sapir, D., Vos, F., Below, R. and Ponserre, S. (2011). "Annual disaster statistical review 2010." Centre for Research on the Epidemiology of Disasters, Belgium.

Haider, R., Rahman, A. A. and Huq, S. (1993). *Cyclone '91: An Environmental and Perceptional Study*. Dhaka, Bangladesh Centre for Advanced Studies.

Hannan, C. (2002). "Mainstreaming gender perspectives in environmental management and mitigation of natural disasters." *Disproportionate Impact of Natural Disasters on Women*, Roundtable Panel and Discussion organized by The United Nations Division for the Advancement of Women and the NGO Committee on the Status of Women in preparation for the 46th Session of the Commission on the Status of Women.

Haque, M. A., Yamamoto, S. S., Mallick, A. A. and Sauerborn, R. (2012a). "Households' perception of climate change and human health risks: A community perspective." *Environmental Health* 11(1): 1. www.ejournal.net/content/1 1/1/1, accessed on 27/4/2015.

Horwich, G. (2000). "Economic lessons of the Kobe earthquake." *Economic Development and Cultural Change* 48(3): 521–542.

Hossain, H., Dodge, C. and Abel, F. (1992). *From Crisis to Development: Coping With Disasters in Bangladesh*. Dhaka, University Press.

Hyndman, J. (2008). "Feminism, conflict and disasters in post-tsunami Sri Lanka." *Gender, Technology and Development* 12(1): 101–121.

Ikeda, K. (1995). "Gender differences in human loss and vulnerability in natural disasters: A case study from Bangladesh." *Indian Journal of Gender Studies* 2(2): 171–193.

IPCC. (2001). "Climate change: Impacts, adaptation, and vulnerability." Contribution of Working Group II to the third assessment report of the Intergovernmental Panel on Climate Change, New York, Cambridge University Press.

Johnson, J. and Galea, S. (2009). "Disasters and Population Health." In K.E. Cherry (ed.) *Lifespan Perspectives on Natural Disasters*, London, Springer: 281–326.

Kahn, M. E. (2005). "The death toll from natural disasters: The role of income, geography, and institutions." *Review of Economics and Statistics* 87(2): 271–284.

Karim, M., Castine, S., Brooks, A., Beare, D., Beveridge, M. and Phillips, M. (2014). "Asset or liability? Aquaculture in a natural disaster prone area." *Ocean & Coastal Management* 96: 188–197.

Karim, M. F. and Mimura, N. (2008). "Impacts of climate change and sea-level rise on cyclonic storm surge floods in Bangladesh." *Global Environmental Change* 18(3): 490–500.

Keim, M. E. (2006). "Cyclones, tsunamis and human health." *Oceanography* 19(2): 40–49.

Lai, T. I., Shih, F. Y, Chiang, W. C., Shen, S. T. and Chen, W. J. (2003). "Strategies of disaster response in the health care system for tropical cyclones: Experience following Typhoon Nari in Taipei City." *Academic Emergency Medicine* 10(10): 1109–1112.

Liang, Y. and Cao, R. (2014). "Is the health status of female victims poorer than males in the post-disaster reconstruction in China: A comparative study of data on male victims in the first survey and double tracking survey data." *BMC Women's Health* 14(1): 18.

Litman, T. (2006). "Lessons from Katrina and Rita: What major disasters can teach transportation planners." *Journal of Transportation Engineering* 132(1): 11–18.

Lundberg, U., & Parr, D. (2000). "Neurohormonal factors, stress, health and gender" in R.M. Eisler & M. Hersen (eds.) *Handbook of Gender, Culture and Health*. UK, Taylor & Francis, 21–41.

MacDonald, R. (2005). "How women were affected by the tsunami: A perspective from Oxfam." *PLoS Medicine* 2(6): e178.

March, C., Smyth, I. and Mukhopadhyay, M. (1999). *A Guide to Gender-Analysis Frameworks*, GB, Oxfam Publications.

Marx, M. Phalkey, R. & Guha-Sapir, D. (2012). "Integrated health, social, and economic impacts of extreme events: Evidence, methods, and tools." *Global Health Action* 5: 19837. http://dx.doi.org/10.3402/gha.v5i0.19837, accessed on 8/6/2015.

Matin, N. and Taher, M. (2001). "The changing emphasis of disasters in Bangladesh NGOs." *Disasters* 25(3): 227–239.

Momsen, J. (2010). *Gender and Development*, New York, Routledge.

Monk, J. and Hanson, S. (2008). !On not excluding half of the human in human geography." *Geographic Thought: A Praxis Perspective* 34(1): 35.

Morrow, B. H. and Enarson, E. (1994). "Making the case for gendered disaster research." Paper presented to the Thirteenth World Congress of Sociology, Bielefeld, Germany.

Morrow, B. H. and Phillips, B. (1999). "What's gender 'got to do with it'?" *International Journal of Mass Emergencies and Disasters* 17(1): 5.

Moser, C. (1993). *Gender Planning and Development: Theory, Practice and Training*, New York, Routledge.

Mukuria, A. G., Aboulafia, C. and Themme, A. (2005). *The Context of Women's Health: Results From the Demographic and Health Surveys, 1994–2001*, USA, ORC Macro.

Nahar, N., Blomstedt, Y., Wu, B., Kandarina, I., Trisnantoro, L. and Kinsman, J. (2014). "Increasing the provision of mental health care for vulnerable, disaster-affected people in Bangladesh." *BMC Public Health* 14(1): 1–9. www.biomedcentral.com/1471-2458/708, accessed on 17/12/2014.

Nasreen, M. (2004). "Disaster research: Exploring sociological approach to disaster in Bangladesh." *Bangladesh e-Journal of Sociology* 1(2): 1–8.

Nelson, V., Meadows, K., Cannon, T., Morton, J. and Martin, A. (2002). "Uncertain predictions, invisible impacts, and the need to mainstream gender in climate change adaptations." *Gender & Development* 10(2): 51–59. doi:10.1080/13552070215911

Neumayer, E. and Plümper, T. (2007). "The gendered nature of natural disasters: The impact of catastrophic events on the gender gap in life expectancy, 1981–2002." *Annals of the Association of American Geographers* 97(3): 551–566.

Nigg, J. M. and Tierney, K. J. (1990). "Explaining differential outcomes in the small business disaster loan application process." Preliminary paper (156), University of Delaware, Disaster Research Center.

Oloruntoba, R. (2010). "An analysis of the Cyclone Larry emergency relief chain: Some key success factors." *International Journal of Production Economics* 126(1): 85–101.

Overstreet, S., Salloum, A., Burch, B. and West, J. (2011). "Challenges associated with childhood exposure to severe natural disasters: Research review and clinical implications." *Journal of Child & Adolescent Trauma* 4(1): 52–68, doi:10.1080/19361521.2011.545103

Oxfam. (2005). "The tsunami's impact on women." www.oxfam.org.uk/what_we_do/issues/conflict_disasters/downloads/bn_tsunami_women.pdf, accessed on 3/5/2005.

Paolisso, M. and Leslie, J. (1995). "Meeting the changing health needs of women in developing countries." *Social Science & Medicine* 40(1): 55–65.

Paul, B. K., Rahman, M. K. and Rakshit, B. C. (2011). "Post-Cyclone Sidr illness patterns in coastal Bangladesh: An empirical study." *Natural Hazards* 56(3): 841–852.

Phalkey, R., Dash, S. R., Mukhopadhyay, A., Runge-Ranzinger, S. and Marx, M. (2012). "Prepared to react? Assessing the functional capacity of the primary health care system in rural Orissa, India to respond to the devastating flood of September 2008." *Global Health Action* 5, doi:10.3402/gha.v5i0.10964

Pittaway, E., Bartolomei, L. and Rees, S. (2007). "Gendered dimensions of the 2004 tsunami and a potential social work response in post-disaster situations." *International Social Work* 50(3): 307–319.

Puentes-Markides, C. (1992). "Women and access to health care." *Social Science & Medicine* 35(4): 619–626.

Rahman, M. (2013). "Climate change, disaster and gender vulnerability: A study on two divisions of Bangladesh." *American Journal of Human Ecology* 2(2): 72–82.

Rashid, S. (2000). "The urban poor in Dhaka city: Their struggles and coping strategies during the floods of 1998." *Disasters* 24(3): 240–253.

Rezwana, N. (2002). "Reproductive health problems and access to healthcare facilities of poor women in Tongi Slums and two villages of Gazipur: A geographical analysis." unpublished MSc Thesis, Dhaka University.

Saroar, M. M. and Routray, J. K. (2012). "Impacts of climatic disasters in coastal Bangladesh: Why does private adaptive capacity differ?" *Regional Environmental Change* 12(1): 169–190.

Sastry, N. and Gregory, J. (2013). "The effect of Hurricane Katrina on the prevalence of health impairments and disability among adults in New Orleans: Differences by age, race, and sex." *Social Science & Medicine* 80: 121–129.

Schmidlin, T. W. (2011). "Public health consequences of the 2008 Hurricane Ike windstorm in Ohio, USA." *Natural Hazards* 58(1): 235–249.

Sen, A. K. (1988). "Family and Food: Sex Bias in Poverty." In P. Bardhan & T.N. Srinivasan (eds.) *Rural Poverty in South Asia*. New York, Columbia University Press.

Sen, A. K. (2002). "Why health equity?" *Health Economics* 11(8): 659–666.

Sen, A. (2004). "How does culture matter." In V. Rao & M. Walton (eds.) *Culture and public action*. USA, The World Bank, 37–58.

Skoufias, E. (2003). "Economic crises and natural disasters: Coping strategies and policy implications." *World Development* 31(7): 1087–1102.

Stover, E. and Vinck, P. (2008). "Cyclone Nargis and the politics of relief and reconstruction aid in Burma (Myanmar)." *JAMA* 300(6): 729–731.

Stromberg, D. (2007). "Natural disasters, economic development, and humanitarian aid." *The Journal of Economic Perspectives* 21(3): 199–222.

Sultana, F. (2009). "Fluid lives: Subjectivities, gender and water in rural Bangladesh." *Gender, Place and Culture* 16(4): 427–444.

Sultana, F. (2010). "Living in hazardous waterscapes: Gendered vulnerabilities and experiences of floods and disasters." *Environmental Hazards* 9(1): 43–53.

Toya, H. and Skidmore, M. (2007). "Economic development and the impacts of natural disasters." *Economics Letters* 94(1): 20–25.

Tunstall, S., Tapsell, S., Green, C., Floyd, P. and George, C. (2006). "The health effects of flooding: Social research results from England and Wales." *Journal of Water and Health* 4: 365–380.

Turner, B. L., Kasperson, R., Matson, P. A., McCarthy, J. J., Corell, L. C., Christensen, L., Eckley, N., Kasperson, J. X., Luers, A., Martello, M. L., Polsky, C., Pulsipher, A. and Schiller, A. (2003). "A framework for vulnerability analysis in sustainability science." *Proceedings of the National Academy of Sciences* 100(14): 8074–8079.

The Watchers. (2015). http://thewatchers.adorraeli.com/2014/09/24/the-statistics-of-natural-disasters-2013-review, accessed on 28/5/2015.Watkins, P. L. and Whaley, D. (2000). "Gender Role Stressors and Women's Health." Eisler, R. M. and Hersen, M. (eds.) *Handbook of Gender, Culture, and Health*, Mahwah, NJ, Lawrence Erlbaum.

Webster, P. J. (2013). "Meteorology: Improve weather forecasts for the developing world." *Nature* 493(7430): 17–19.

Weisman, C. S. (1997). "Changing definitions of women's health: Implications for health care and policy." *Maternal and Child Health Journal* 1(3): 179–189.

West, D. M. and Orr, M. (2007). "Race, gender, and communications in natural disasters." *Policy Studies Journal* 35(4): 569–586.

Wiest, R. E., Mocellin, J. S. P and Motsisi, D. T. (1994). *The Needs of Women in Disasters and Emergencies*, Disaster Research Institute, Canada, University of Manitoba.

Zilversmit, L., Sappenfield, O., Zotti, M. and McGehee, M. A. (2014). "Preparedness planning for emergencies among postpartum women in Arkansas during 2009." *Women's Health Issues* 24(1): e83–e88.

2   Research methodology and fieldwork in Bangladesh

Our reading of the field suggests the best theoretical work with the most urgently needed practical dimension is written from the world's most dangerous places and about women and men who learn to live with risk.
—(Enarson and Meyreles 2004, p. 74)

2.1 Introduction

I have a selected qualitative research methodology to explore in detail the conditions prevailing in the study area and in an effort to reveal the factors shaping men's and women's access to healthcare during disaster. Barguna, one of the coastal districts of Bangladesh, is the study area of my research. Barguna is vulnerable to cyclone disasters because of its location, lack of building infrastructure, economic conditions, insufficient transport links and, above all, its poor healthcare facilities. It is an area where marked gender differences in socio-economic status are found among the inhabitants. In the following sections, firstly I will discuss the research design of my study and then follow on to a description of the study area and fieldwork. I will also share my experience of how the fieldwork became a great challenge for me, as a female researcher, to reach and stay in a remote coastal region of Bangladesh.

2.2 Research questions

Inspired by findings from the literature review (formal theory) and my own existing understanding of the field (tacit theory) (Marshall and Rossman 1999, p. 29 & 38), I formulated the intellectual puzzles and research questions for the research (Mason 2002, p. 18) as follows.

Causal puzzle

What influence do disasters have on health conditions and accessibility to healthcare with regard to gender?

Mechanical puzzle

How do prevailing policies and practices regarding the healthcare facilities for pre-, during and post-disaster periods take account of the gendered effects of natural disasters?

Expressing these as puzzles helped me to formulate the set of research questions which is designed to address the essence of enquiry in the sense of their ontology and epistemology. According to Blaikie, 'a research project is built on the foundation of its research questions' (Blaikie 2009, p. 58) which 'should be general enough to permit exploration but focused enough to delimit the study'(Marshall and Rossman 1999, p. 38).

The questions of my research are as follow:

1   What are the impacts of disasters on health conditions in the coastal region of Bangladesh?
2   What are the gendered impacts of disasters on health conditions in the coastal region of Bangladesh? What are the factors affecting the health conditions of men and women in disasters?
3   What are the impacts of disasters on access to health care facilities?
4   What are the gendered impacts of disasters on access to healthcare facilities in the coastal region of Bangladesh? What are the factors affecting the healthcare access of men and women in disasters?
5   What are the current policies and practices regarding pre- and post-disaster healthcare provision, and to what extent do they take account of gender?
6   What improvements might be made to pre- and post-disaster healthcare provision for women?

2.3 Qualitative research and research strategy

The intellectual puzzles, research questions and aims of the research call for a detailed, in-depth investigation of the interplay of factors, power relations and social processes influencing the impacts of disasters on health conditions and healthcare access in different social contexts. To pursue this enquiry, qualitative research is 'uniquely suited to uncovering the unexpected and exploring new avenues' (Marshall and Rossman 1999, p. 38), and qualitative methods are well established in the discipline of Geography, although they are seldom used for this topic, especially within Bangladesh.

Through qualitative research, 'a wide array of dimensions of the social world' can be explored, 'including the texture and weave of everyday life, the understandings, experiences and imaginings' of research participants and 'the ways that social processes, institutions, discourses or relationships work' and influence people's lives and different decisions (Mason 2002, p. 1). Qualitative research, which is 'more sympathetic to the human beings' (Lowe and Short 1990, p. 7), helped me to investigate how disasters influence people's decisions and practices in regards of healthcare. Again, being 'pragmatic, interpretive, and grounded in the lived experiences of people' (Marshall and Rossman 1999, p. 2), qualitative research 'has much to offer' to enrich knowledge of health and healthcare (Mays and Pope 2000, p. 52) and is gaining greater importance in health and health service research (Mays and Pope 2000). I therefore approached fieldwork in the study area with an inductive research strategy (Blaikie 2009, p. 102) and flexible research design (Robson 2002, p. 86).

2.4 Approaches and underlying philosophy

My research is grounded in both anti-naturalist and hermeneutical philosophies. 'Hermeneutics' is the study of interpretation and meaning or understanding (Graham 2005; Hoggart et al. 2002) and 'offers a very different perspective on truth from empiricism, positivism and critical realism, philosophies that all distinguished sharply between fact and fiction' (Hoggart et al. 2002, p. 25). These philosophies helped me to work adopting the position of learner rather than expert by following a 'bottom up' method (Blaikie 2009, p. 139) and to understand and interpret visible and invisible factors and processes that lay behind individuals' decisions in the pre- during- post-disaster periods. The research design is situated within interpretivist and humanist approaches. Both approaches see people as the primary data source. Humanist approaches are concerned with 'people as a social actors or active social agents' (Mason 2002, p. 56) and interpretivist approaches focus on people with 'understanding the social world people have produced and which they reproduce through their continuing activities' (Blaikie 2009, p. 115). As my research aims to reveal the health impacts of disasters and what people do during disasters in regards of their health care problems, and explores the factors modifying their decisions in seeking health care facilities, these approaches are most appropriate. They allow much-needed ground-up understandings of the phenomena being investigated, seeking 'people's individual and collective understandings, reasoning processes, social norms' with an 'insider view' not with an 'outsider view' (Mason 2002, p. 56 and Blaikie 2009).

2.5 A framework for the research design

According to Robson, research design concerns 'the various things which should be thought and kept in mind when carrying out a research project' (Robson 2002, p. 80). I followed Mason's advice to produce a detailed 'internal' research design for the researcher's own use and to practice strategic thinking and reflection from the start, but not taking this design as 'once-and-for-all' blueprint for the research (Mason 2002, p. 25). In my research design, all the components (mentioned below) were repeatedly revisited as the research progressed. Here, the components 'purpose', 'theory' and 'research questions' helped me to choose 'methods' and 'sampling strategy' with confidence (Blaikie 2009). During and after the field study and collecting and analyzing data, these components have been renewed (highlighted texts shows the changes) within the flexible research design (Robson 2002).

Purpose	Revealing the gendered impacts of different disasters on health and healthcare access
Theory	-Disasters have uneven spatial and gendered impacts
	-Poverty has a strong relation with disaster impacts
	-Natural disasters have more impacts on women
	-Women need special healthcare in disasters
Research questions	-Mentioned in section 2.2
Methods	-One-to-one in-depth interviews
	-Focus group discussion
	-Participatory diagramming
	-Participatory mapping (deducted)
	-Observation
	-Round table discussions (deducted)
	-Seminar and open discussions with professionals **(added)**
Sampling strategy	-Key informants from the study area
	-Inhabitants of the study area
	-Healthcare staff/Doctors **(Changed to doctors)**
	-Government/NGO professionals/Practitioners
	-Records of healthcare centres
	-Planning and policy documents of both government and NGOs

2.6 The study area

Before entering a detailed discussion on data collection, a short description of the study area will be presented in the following sections.

Bangladesh as a disaster-prone country

Bangladesh is one of the most disaster-prone countries in the world (Matin and Taher 2001). It is situated at the crossing point of the Himalayas to the North and the Bay of Bengal to the South, with fertile, flat land in between (Rashid 1991). This location has given the country life-giving monsoons on one hand, and catastrophic disasters like tropical cyclones, storm surges, floods, droughts and erosion on the other (Ali 1999). Bangladesh experienced 234 natural disasters from 1980 to 2010, which killed 191,836 people and affected about 323,480,264 others. Among the disasters, cyclones were the major disasters in both the number of occurrences and casualties, making the coastal districts the most vulnerable region in the country (PreventionWeb 2014).

Bangladesh is prone to at least one major tropical cyclone every year (in probabilistic terms) (Mooley 1980). According to the records of the last 200 years, the coastal belt has faced 70 major cyclones. About 900,000 people have died during the last 35 years alone due to these catastrophes (Islam and Ahmad 2004). Among these cyclones, Bhola Cyclone was the most devastating and struck on November 12, 1970 (Islam et al. 2011). More than 500,000 people lost their lives in this storm, which was equivalent to a Category 3 hurricane. The most affected area was Tazumuddin, which lost 45% of its population of 167,000 in one night (Paul and Rahman 2006). Another cyclone on 29–30 April, 1991 was particularly severe, causing widespread damage of US$2.07 billion and taking another

138,882 lives (Bern et al. 1993). In recent years, Cyclone Sidr on 15 November, 2007 triggered giant waves up to 30ft (7m) high and killed more than 10,000 people (although, according to government records, 3,400 people were killed, Government 2008) in the south western coastal belt of Bangladesh. The damage was about US$5.3 billion (Hossain et al. 2008). Beside cyclones (accompanied by tidal surges), the major natural hazards in Bangladesh are floods, river bank erosion and the arsenic contamination of ground water (Matin and Taher 2001). There is also a high vulnerability to climate change.

Bangladesh's geographical location, large population and socio-economic conditions have also created great vulnerability to all of the above hazards. The total population of Bangladesh was over 142.3 million people, and the density was 964 per sq. km in 2011 (BBS 2011). More than one-third of the people live in poverty: the poverty-headcount ratio at $1.25 a day (PPP) was 43.3% in 2010, and income distribution is very unequal (World Bank 2015). Agriculture contributes about 20.24% to the total Gross Domestic Product (GDP) and occupies about 48.1% of the labour force of the country as the single largest producing sector of the economy (BBS 2011). This dependency makes the economy highly sensitive to climate change (Agrawala et al. 2003). Practically, in this disaster-prone country with a high density of population and poor economic conditions, no natural hazard occurs without adversely affecting a large number of people and their assets and even influencing the whole country beyond the disaster-struck area (Matin and Taher 2001; Loayza et al. 2009).

Beside its economic condition, Bangladesh has increased vulnerability because of low health status and inadequate healthcare facilities, particularly in rural areas (Baden et al. 1994; Mahmood et al. 2010; Ahmed et al. 2011). In this country, almost half of its inhabitants (women) have the highest energy deficiency in the world (del Nino et al. 2001 cited in Cannon 2002), which reduces women's capacity to cope with the effects of a hazard (Cannon 2002). Different remedial methods are practised in Bangladesh, ranging from indigenous self-care methods to folk and modern medicine, though treatment options are importantly determined by urban-rural location and social factors like poverty and gender (Cannon 2002). Women in this country seek healthcare for illness significantly less often than men (Ahmed 2000) or receive poorer healthcare (Cannon 2002). Only 18% of women give birth to their children in the presence of a doctor or other health professional (Prata et al. 2012), and about 87% of births take place at home (this is even higher in the rural areas). In addition, more women die from pregnancy-related complications in rural areas (BBS 2011). So, locational discrimination in the health sector is present in the rural areas and even in the remote urban areas of Bangladesh, which is intensified in the highly vulnerable coastal regions during cyclone disasters. Barguna is one of these highly vulnerable regions, sharing the part of the southern coastal line of Bangladesh with the Bay of Bengal.

Barguna: a coastal district in Bangladesh

Barguna is a densely populated remote coastal district with 950,737 people in 1830 sq.km area. It is well known to the world as a Sidr- and Aila-affected area.

> People of this area are mostly poor. They depend on cultivation, fishing and small business for their livelihood. Road communication is not good here because two large rivers (Payra and Bishkhali) divide the district into three parts. Diarrhoea is endemic in this area, especially in Spring and Summer. An acute shortage of doctors (33 out of 141 sanctioned posts) makes better health delivery difficult. Natural disasters make the scenery troublesome. Load shedding is a common event due to this situation. We frequently pass three to five days without electricity due to rain and thunder shower leading to load shedding. Hence we cannot perform our digital office activities (use of computers).
>
> —(Health Bulletin 2013)

This quotation reflects the difficult conditions in Barguna, as well as the wariness of the main local Government health authority about disaster health impacts. Due to the poor socio-economic conditions and healthcare provision in Barguna, health conditions become worse during even common natural hazards, and intensify during disasters like cyclones. Barguna is often in the path of cyclones – it was affected by severe cyclones in 1935, 1965, 1970 and, more recently, Sidr in 2007 (Tamima 2009) and Cyclone Mahasen in 2013. A brief description of Cyclone Sidr and Cyclone Mahasen follows.

Cyclone Sidr

Cyclone Sidr was first observed on 9 November, 2007 as a weak, low-level circulation near the Nicobar Islands. By 15 November, 2007, this depression turned into a Category 4 storm and made landfall across the Barisal Coast, with sustained winds of up to 240 km per hour and a storm surge of 5.5 to 6 metres at the mouth of Baleswar River. In the days before Cyclone Sidr hit the country, the Bangladesh Meteorological Department (BMD) issued cyclone warnings and advisory messages using the Government's warning signal system. Warnings were sent to communities frequently, and 44,000 volunteers activated community-based warning systems, utilizing megaphones and other devices. Cyclone Sidr affected 12 districts in south western Bangladesh, killed 3,400 people and injured 55,000 people (Government 2008). During Sidr, Barguna district lost 1,335 inhabitants (about 40% of the nationwide casualties), 60–70% of crops and 95,412 houses, which were either fully or partially damaged (NIRAPAD 2007 cited in Tamnia 2009).

Cyclone Mahasen

Cyclone Mahasen formed on 11 May, 2013 in the Bay of Bengal and struck the south east coast of Bangladesh (CDMP 2014), affecting some areas of nine coastal districts. This cyclone claimed 17 lives, injured 5,374 and affected about 1.5 million people (DDM 2013) in the coastal region. Barguna was one of the worst affected coastal districts. In Barguna district six upazilas, 38 unions and 60,000 people were affected. About 6,856 houses were completely damaged.

Seven deaths were reported (CDMP 2014). According to the government report, the impact of the cyclone was less than anticipated because of the good preparations and the reduced speed of the cyclone (DDM 2013).

Besides yearly cyclones, coastal flooding and salinity on cultivable land (Miah et al. 2004) are significant natural hazards in Barguna. Again, like other coastal districts of Bangladesh, Barguna is very vulnerable to climate change effects (Ali 1999), which include sea level rise (Ali 1996 and Agrawala et al. 2003), high temperatures, changing seasonal precipitation (Agrawala et al. 2003), increase in flood risk areas and depth of flooding (Karim and Mimura 2008) and an increase in cyclone intensity (Ali 1996; Agrawala et al. 2003). Barguna is not prepared for these serious impacts of climate change.

With regard to socio-economic conditions, only 10.32% of Barguna's population live in urban areas (BBS 2006). A large number of people in this district are landless (about 85% of the total), and agriculture and fisheries are the main economic activities. The literacy rate is on average only 55.3% and much lower among the female inhabitants (Byuro 2006).

Healthcare centres in Barguna district

Government and private healthcare facilities are both available in Barguna district, though private healthcare facilities are still at an early stage. Among the government facilities, there is one District Hospital, four Upazila Health Complexes, eight Union Sub-Centres, 26 Union Health and Family Welfare Centres, one Mother and Child Welfare Centre and 108 community clinics (Health Bulletin 2013). Among these centres, the district hospital offers 100 beds, the Upazila Health Complexes offer 211 beds and the Mother and Child Welfare Centre offers ten beds. About 141 doctors' posts are sanctioned (permitted by the government) in these centres (excluding community clinics), which means that the ratio between patient and doctor is one doctor for 6,743 people in Barguna. However, in practical terms, due to vacant posts, the ratio has seen an alarming downturn to one doctor for 28,810 people, despite the large number of patients visiting these government centres for treatment. According to the Health Bulletin 2013, between January and December in 2012, about 932,337 patients visited the outpatient departments (OPD) of these centres, 7,734 patients visited the emergency departments and about 20,764 people were admitted as inpatients. According to the Health Bulletin 2013, about 23 private clinics or facilities are located in Barguna district, but they offer only 30 beds in total. Therefore, overall, healthcare facilities in Barguna district are insufficient.

Study areas in Barguna district: Barguna municipality and Tentulbaria village

Barguna district is divided into several administrative units, including five Upazilas, 38 Unions, 560 villages and four municipalities. To conduct my fieldwork,

I selected Barguna Paurashava (municipality) and Tentulbaria village of Barguna Sadar Upazila in Barguna district (Map 2.1). These two areas were selected as typical of an urban area and a remote rural area within Barguna district. The logic behind this is to focus on the importance of location within the research and to reveal the differences between urban and rural areas in regards of health and healthcare provision during disasters.

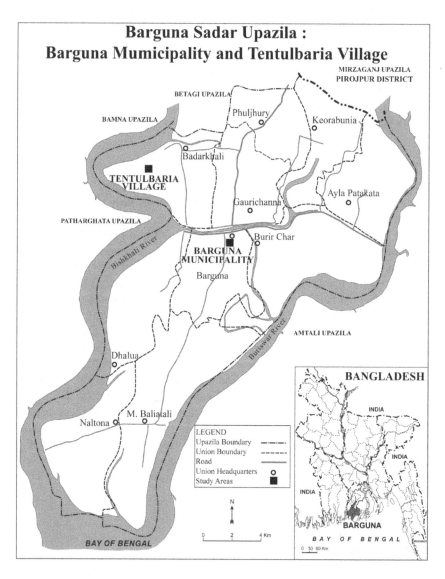

Map 2.1 Barguna Sadar Upazila: Barguna municipality and Tentulbaria village

Barguna Paurashava

The first study area, Barguna Paurashava (municipality), is one of the south-ernmost municipalities in Bangladesh (LGED 2015). It is the only urban area in the Barguna Sadar Upazila, with a small number of buildings (only 10.78% of the total households in the municipality), one highway and some local roads and a minimum of urban facilities (only 1.27 % of the population have access to tap water, and 56.49% of urban households have an electricity connection, BBS 2006). Local deep-tube wells are the main source of drinking water, though due to the presence of dissolved iron, this water looks reddish-yellow in colour. Local inhabitants usually avoid this water for cooking, preferring rainwater and pond water. Local shops are the only place for shopping, as there are no brand shops or shopping complexes in this municipality. There are five colleges, but few of them have post- graduate study options. Students have to go to another district, Barisal, to do master's degrees, which is a constraint for them, especially female students.

Remoteness of the area

Barguna municipality is 319km from the capital, but its remoteness is intensified because of its poor transport system and lack of other modern facilities – good internet connections, modern electronics, packed or tinned foods, cyber cafes and so on are not easily available in this area. Even daily newspapers reach Barguna the following evening. Travelling by waterway is the safest way to get to Barguna, but it is very time consuming. There is also a road network, but it also requires fer-ries to cross the rivers, and the facilities are not well managed. Barguna munici-pality has lower socio-economic development than the rest of the country. The literacy rate is lower than the national average. Only 56.97% of girls are sent to school, but this rate drastically decreases to 15.12% after the age of 19 years. Girls have arranged marriages at a very young age, some at ten years (BBS 2006). A trend has been noticed in this urban area among some of the wealthier older men of having two or more wives at a time, which is rare in other urban areas of Bangladesh. This tradition is accepted as normal in Barguna society, women see-ing it as their bad luck.

The society is still very patriarchal. When I stayed in Barguna for the field-work, I witnessed firsthand that women were treated as second-class citizens. I was invited to their homes and spent hours or days with them. They are always dependent on male members of the family – husband, father, senior or even junior brother, son – for economic support and even simple decisions. Though it seems that these women are becoming aware of alternative ways of living, they cannot take decisions, not even on their healthcare access, because of a combination of lack of freedom and lack of confidence. Sometimes, it is older women themselves who perpetuate this situation for junior female members of the family.

However, vulnerability of local women to health problems increases due to insufficient healthcare facilities in the Barguna municipality. Short descriptions of these facilities are made in the following paragraphs.

Healthcare facilities in Barguna municipality

The Government Hospital is the major healthcare centre for the region, located in the Barguna municipality (Figure 2.1). This healthcare centre is open 24 hours a day and delivers emergency services and, above all, admissions for operations and checkups are free or provided at a very low cost. Most of the poor respondents in my research are highly dependent on this hospital. However, this 100-bed hospital had only 50 beds operational; only nine doctors against 21 sanctioned posts with no female doctor; and 23 nurses against 37 sanctioned posts in 2012 (Health Bulletin 2013). According to the Health Bulletin (2013), about 89,284 patients visited the outpatient department, 3,859 patients visited the emergency department and about 12,463 patients were admitted to this hospital in 2012.

About ten private clinics, including NGO healthcare facilities, are located in Barguna municipality (information from Municipality Office in 2013). However, private clinics are very costly and do not provide any emergency services. Most government doctors and other MBBS doctors in Barguna practise privately in these clinics. Even a few doctors from Barisal and Dhaka visit these private clinics periodically.

Again, there are private chambers of healthcare providers in Barguna municipality. These chambers and pharmacies are mostly located in the main bazaar of Barguna municipality, which is called 'Pharmacy Patti'. A range of practitioners,

Figure 2.1 Government Hospital: Barguna municipality

Source: Fieldwork, 2013

from MBBS doctors to non-qualified medical providers, are found to be practising in the bazaar. Many MBBS doctors in Barguna (12–13 in number) practise in these chambers, though non-qualified medical providers are thought to be higher in number (about 30–40 according to the local inhabitants). The costs of visits to the doctor vary from 300 taka to 10 taka (£2.30 to 0.08 pence) according to their qualifications and popularity. In many cases, people also collect their medicines without consulting a doctor, because medicines for the most common diseases are known to them, and, sometimes, they want to avoid the cost of a doctor.

Tentulbaria village

Tentulbaria is one of the villages of Badarkhali Union, Barguna Sadar Upazila, Barguna (Map 2.1). This village is just beside the river Bishkhali and was severely affected during Cyclone Sidr. It remains highly vulnerable. The embankment beside the river is still damaged. The main problem of this village is the poor transport system. Travelling on waterways by local trawler is the only way to reach the village, and this transport is poorly managed. River ports are unfriendly to passengers, especially women, children, older and disabled people.

There are about 732 households and 3,300 people living in this village. Most of the houses are kutcha (made of tin and wood) and a small number are semi pucka (brick built). Both types of house are unsafe in strong winds and storm water during cyclones. With regard to the main activities of the households, 34.42% make a living from agriculture and forestry, 29.51% from fishing, 11.38% from agricultural labour and 9.29% from business (BBS 2006). A tube well is the main source of water in this village, and less than half of the households have electricity (BBS 2006).

More boys attend the school than girls, and the number of school leavers is very high after the age of 15 for both groups, though a higher rate is found among girls. Girls have arranged marriages at a very early age, as young as ten to 15. Most women do household work and make up 43% of the total population of the village (BBS 2006).

This village has only one school, and there is no post office, no police station, no hotel or restaurant.

Healthcare facilities for Tentulbaria village

There is no healthcare centre or pharmacy in Tentulbaria village. Only the Government EPI programme works periodically at the primary school for the mothers and children of the village. Villagers have to depend on the pharmacies at nearby villages to collect common medicines and on the healthcare centres of Barguna Sadar Upazila for their treatment. Barguna Sadar Upazila has one Upazila Health Complex, three Union Sub-Centres, seven Union Health and Family Welfare Centres, one Mother and Child Welfare Centre and 33 community clinics. Among all these Government health centres, only the Mother and Child Welfare Centre offers a ten-bed facility. There are 12 sanctioned doctors posted in these centres, but only two doctors' posts were filled in 2013 (Health

Bulletin 2014). According to the Health Bulletin (2014), there are four private clinics and two NGO clinics located in the Barguna Sadar Upazila. About 580 patients visited the outpatient department (OPD) of the private clinic, 1,489 patients were admitted in 2013 and 20,401 patients visited the NGO clinics. But Tentulbaria villagers mostly have to travel to Barguna municipality for health-care due to the absence of emergency services and lack of treatment facilities in these centres. For common and minor health problems, they mostly visit Phuljhuri Union Sub-Centers. One SACMO (Sub Assistant Medical Officer) and one FWV (Family Welfare Visitor) are working in this centre. Beside the other services, South Asia Partnership (SAP) Bangladesh has an ongoing pro-gramme at this centre, which monitors child growth and gives food support to lactating mothers (information provided by the SAP Bangladesh office in 2013). Mothers can receive services such as maternal healthcare, birth control and child care at this centre. However, at present, this centre offers facilities for normal deliveries, but critical deliveries need to be transferred to Barguna municipality (information provided by the Plan International Bangladesh in 2013). Due to the lack of proper facilities, most of the time emergencies and serious illnesses require visits to Barguna municipality.

2.7 Healthcare facilities in Barguna: a summary

Lack of proper healthcare facilities is a major problem in Barguna. The very small numbers of specialist doctors, insufficient number of pathology laboratories, lack of beds in the district hospital (100 beds for 261,363 people) and the presence of a few expensive, private healthcare centres have created very poor healthcare con-ditions in Barguna. The private sector is not able to provide all of the necessary treatments but only offers better facilities and common treatments for richer peo-ple. These people can visit other districts if required. However, problems become intensified for the poor because they have to depend on the government sector, which does not provide full free treatment and refer the patients to other centres. Additionally, mismanagement, misbehaviour and corruption are also big prob-lems of this sector (mentioned by most of the respondents). These poor inhabitants cannot afford costly private ambulances and have to borrow money to complete their treatment. Again, the unequal distribution of healthcare centres has also cre-ated vulnerability to health problems for the remote areas like Tentulbaria village. Even travelling to the municipality from their remote locations does not always guarantee the necessary healthcare. Most of the emergency cases are referred to Barisal district, but it is not possible for everyone to afford the treatment cost. So, Barguna inhabitants face inadequate treatment facilities, raising questions over poorer residents' access, especially poor women's access, to healthcare. Though there has been an improvement with regard to women's access to healthcare facili-ties for immunizations during their prenatal period, differences are still noticeable among the different economic groups. Most of the women are dependent on the government EPI programme for their immunization, but solvent women can visit the NGO-based private clinics for this purpose. Again, home births are common in Barguna because of the low cost, but women from better-off families can afford trained birth attendants during childbirth or visit private clinics for emergency caesarean operations.

The above discussion has identified the vulnerability of Barguna as a disaster-prone region, introduced the study areas and highlighted the absence of good-quality, affordable healthcare facilities for all. These conditions lead to the aims of the research reported here and demonstrate why Barguna has been chosen as the study area. In the next sections, I go on to describe the fieldwork in Barguna, the methodology and methods used to collect data and, of course, the great challenges I faced as a female researcher in this traditional society.

2.8 Fieldwork in the coastal region of Bangladesh: data collection and challenges faced in the field

I started my journey from England to Bangladesh with great hopes and enthusi-asm to conduct successful fieldwork. On reaching Bangladesh, my first plan was to collect background information on Barguna from the capital city Dhaka. For this, I visited the Bangladesh Bureau of Statistics (BBS) and Disaster Research Training and Management Centre (DRTMC) and collected information (data and local maps) on Barguna District. This information helped me to get to know the administrative units, socio-economic conditions of the inhabitants and disaster-prone areas of Barguna. I could have focused on the local urban and rural areas of Barguna more precisely. I had also planned to collect more information on the healthcare services and organizations working in Barguna, but doing so in Dhaka proved to be difficult and time consuming. So, I had to leave this plan to collect more background information from the local office of Barguna district.

Arrival in Barguna and adaptation

After collecting information from the Dhaka offices, I set off for Barguna. A jour-ney by steamer from Dhaka to Barguna is the safest method of travel, but it is very time consuming – it took 27 hours. Due to bad weather, mismanagement and corruption within the steamer authority, the normal 19–20-hour journey needed an extra six to seven hours. The steamer was overloaded with passengers and business products, exceeding its capacity, and was steered by a co-captain, which caused an unexpected incident during the journey. The steamer got stuck on a river island and had to be rescued by another ship.

After a long journey, I reached Barguna and was received by my key informant, who is also my relative. I stayed at his home for few days before moving to my rented house. I had planned to stay at Barguna municipality for the whole period of fieldwork. There were two reasons for this. First of all, I wanted to select this municipality as my urban study area, because the urban facilities that Barguna dis-trict has are mostly concentrated in this area, including most of the head offices of administration, health and local NGOs. A second reason was that houses for rent were only available in this municipality. There are very few local hotels, and they are not safe due to the social pressures on staying alone as a female researcher, and so I had planned to rent a house on a short-term contract. I had also requested my senior relatives (both my parents-in-law) to stay with me for these months of fieldwork. This would enable easy access to Barguna society with respect and transparency. Often, the very first question from respondents was where am I staying and with whom. Women are not expected to live alone in Barguna and always attract unwanted curiosity. I was believed and respected (as respondents mentioned several times with compliments) because I was staying and doing my research following the prevailing social rules in Barguna. This greatly helped in gaining access for in-depth interviews in this traditional society. After moving to the unfurnished rented house, I needed some days to make it liveable for the next few months. This also allowed me the chance to visit local bazaars and talk with local inhabitants. It helped me to observe local men and women and their daily activities, social behaviour and problems faced, which I discuss further in Chapters 3, 4, 5 and 6 of this book.

There are many problems in Barguna municipality. But above all, collecting cooking water from the tube well is a very important task. Most women have to collect the drinking water from the roadside tube wells, whereas women from the much affluent families are not allowed to do. So, they have to recruit other women, especially poor women, to help them, which I had to follow, too. Besides this, it is not accepted for any woman (especially young women) to do shopping or anything else on their own in Barguna. They need company, either a male or senior female. This social attitude also became a challenge for me. I had to receive help from one of my male key informants (my relative) to visit different areas of Barguna. I was fortunate that he was willing to help my fieldwork. Here it should be mentioned that a few professional women from other areas of Bangladesh also work in Barguna, but they are not beyond curiosity and questions, as they mentioned in their interviews. Their affiliation with local institutions is a great support for them. But for a researcher like me, with a plan to do in-depth interviews at the local level, the culture was quite challenging. I had to answer questions several times about my identity and relation with my male informant. These initial experiences helped me to understand the social attitude and rules of Barguna society.

Again, women in Barguna mostly wear a hijab (head scarf, sometimes covering their face) and Borka (extra cover on the dress). I also changed my usual dress to avoid unwanted attention from the local people and to allow me free movement to work in public places. Sometimes, I found that the male inhabitants also did not feel free if I did not wear the usual dress of Barguna women.

Preparation for data collection

I took several days to prepare before starting data collection. Reaching Barguna, I found that local people are even more conservative than I anticipated. They, especially women, do not feel free to talk with unknown people. I felt that interviewing them without having any introductory meeting would not allow me to fulfil my aims in data collection. So, at first, I introduced myself as a relative of my key respondent and talked with them on different topics before mentioning the interviews. I found that talking with them about family life, daily household works, parenting and my life in Britain made them feel comfortable with me. So, I spent several days talking with people of different socio-economic statuses and occupations, and, through these meetings, I began to create a framework in which to start the fieldwork.

Methods and sampling strategies

This research employs grounded theorizing, where 'explanation and theory are fashioned directly from the emerging analysis of the data using a constant comparative method' (Glaser and Strauss 1967 cited in Mason 2002, p. 180), and the grounded theory method, in which the 'researcher develops inductive theoretical analyses from their collected data and subsequently gathers further data to check these analyses' (Charmaz and Bryant 2011, p. 292). For this, I selected different methods and theoretical sampling strategies to gather data. The methods are one-to-one in-depth-interviews, focus group discussions, participatory diagramming, observation, seminar and open discussion with professionals. Consequently, I conducted data collections and analysis and formed my theoretical categories, until I felt that the categories had reached saturation. This involved four months of fieldwork in Bangladesh. Table 2.1 summarises the methods conducted. I go on to describe how each one was implemented in the field.

Table 2.1 Methods conducted in the fieldwork 2013

Methods	*Respondents/ participants*	*Location*	*Category*	*Number of respondents/ events*
Interviews	Inhabitants	**Barguna Municipality**	Household work	3 – Female (solvent and poor economic family status)
			Labour	3- Female
			Businessman (well established and small/ part time labour)	2-Male
			Service	6-Male and Female (high officials to contract service holders)
			Student	3 – Male and Female
			Religious	1-Male
			Total	18 (8 Male & 11 Female)
		Tentulbaria Village	Household work	3 – Female
			Agriculture/ fishery	2 – Male

(Continued)

Table 2.1 (Continued)

Methods	Respondents/ participants	Location	Category	Number of respondents/ events
			Labour	1-Male
			Businessman (well established and small)	2-Male
			Service – local school teacher	2- Male and Female
			Dependent	1 – Female
			Student	1 -Male
			Total	12 (7 Male & 5 Female)
	Total 30 Individuals (14 Male & 16 Female)			
	NGOs		Works on disaster, response programmes, healthcare facilities, gender	**9**
	Doctors		Government Hospital Healthcare centres	**2**
	Mayor of the municipality	Barguna Municipality		**1**
	Chairman of the Union	Badarkhali Union		**1**
Focus group discussions & participatory diagramming	Inhabitants	Barguna Municipality	Male Female	**2 Events** (4–6 participants)
		Tentulbaria Village	Male Female	**2 Events** (4–7 Participants)
Observation		Barguna Municipality	Healthcare centre Cyclone Shelter Vulnerable residential areas	
		Tentulbaria Village	Cyclone Shelter Vulnerable residential areas	
Seminar & open discussions professionals	Professionals	DRTMC, Dhaka University	Works on Gender, Disaster Managements	**1**

Selecting respondents for interviews and focus group discussions

To select my respondents, I used key informant techniques, snowballing and field observation methods using theoretical sampling. As Houston describes it, 'the key informant technique is assessed in terms of the ability of different types of informants to report on various aspects of the social system in which they perform role functions' (Houston and Sudman 1975, p. 151). My first key informant and relative, who is a well-established businessman in Barguna, introduced me to six further key informants of different occupations and genders. They are one senior ex-member of the Union Parishad, one senior school teacher, one family planning fieldworker who is very familiar with the local people, one final year undergraduate student, one housewife and one NGO professional. With the help of these key informants and through the snowballing method, I have chosen respondents from the local inhabitants and officials following theoretical sampling strategy. Theoretical sampling 'is concerned with constructing a sample (sometimes called a study group) which is meaningful theoretically and empirically, because it builds in certain characteristics or criteria which help to develop and test' the theory or argument (Mason 2002, p. 124). In this regard, I have chosen two groups of respondents, which was required to develop the properties of a tentative category (Charmaz and Bryant 2011 in Silverman 2011, p. 292) and investigate the research questions of my study. One of the groups is local inhabitants from Barguna municipality and Tentulbaria village who shared their experiences of cyclones in regards of health and healthcare access; and the other group is government officials, such as the mayor of the municipality, chairman (Union Parishad), doctors and NGO workers, who talked about the prevailing disaster management plans and policies and their experiences in responding to disasters. From studying documents and literature and observing the field, I chose occupation to define my sampling categories, which are also cross-cut by gender, education and economic conditions. The target characteristics for respondents were men and women aged over 18 who have experience of disasters and who were from a range of occupational statuses, such as fisherman, farmer, housewife, service holder and student, also including from different economic status and educational qualifications. The target characteristics for the official respondents were men and women who are working on health and disasters and affiliated with several institutions.

Pilot survey and interviews

Before embarking on the in-depth interviews, I conducted a pilot survey to test whether the proposed interview schedule was suited to its purpose. The first interviewee of the pilot survey was an educated housewife who lives in the Barguna Paurashava (municipality). The semi-structured interview schedule was found to work well, and I added only a couple of topics to it following the pilot. I recorded the interview, with her permission. After completing the interview, I checked the interview and edited the schedule and then did the second interview with the edited schedule. The second respondent was a male of 20 years who is a salesman in a local shop and a part-time student. This interview showed me that the interview schedule did not need any further changes, and I then finalized it.

In-depth interviewing

In-depth interviews are 'a conversation with purpose' (Kahn and Cannell 1957, p. 149 cited in Marshall and Rossman 1999, p. 108), which 'give an authentic insight into people's experiences' (Silverman 1993, p. 91). They are the major data source and method in my research to explore the 'subjective values, beliefs and thoughts of the individual respondent' on gendered disaster impacts on health and health-care accessibility (Valentine 2005, p. 112). My interview schedule included a list of topic headings and key questions under these headings, including personal experiences of disasters, conditions of health and healthcare accessibility during and after disasters, factors affecting healthcare accessibility and differences in regards to individual, gender and socio-economic status. The language used was Bangla, the native language in Barguna. With permission, all interviews were digitally recorded and notes taken. All the interviews were transcribed later for data analysis.

It took almost one hour to complete each interview, though I spent more time introducing myself and my research to the interviewee. To become familiar with them and conduct the interviews, I visited their houses and offices. Sitting with them in their own environment helped me to observe them and understand their situation better. Several invitations to their houses allowed me to be in the kitchen with them, have meals and even spend nights at their houses. Many of them shared personal opinions, private incidents and their emotional responses to issues of health, healthcare access and disasters. I was happy that even male respondents spontaneously shared their opinions with me, which I had been concerned would not happen before starting the fieldwork.

Interviewing respondents among the local inhabitants:
Barguna municipality

I conducted a total 18 in-depth interviews with eight male and 11 female residents of Barguna municipality (Table 2.1). To choose the respondents, I have tried to select both males and females in the same numbers, but categories like household work, business, religious do not include both males and females. Among service holders, I selected officials who are working in government offices, NGOs, shops and healthcare centres.

Interviewing respondents among the local inhabitants:
Tentulbaria village

The most challenging part of my fieldwork was to reach Tentulbaria village. There was no other way except by waterway to visit this village, and local boats were the only vehicles available. These are locally called trawlers, each with a helmsman and a motor engine. The boats do not have any safety measures, such as life jackets, and no seats to sit on. These boats become very vulnerable in storms. Unfortunately, I was in Barguna during the rainy season, and there were local thunderstorms with strong winds every day. As I do not know how to swim,

I could not risk taking the full trawler journey. So, I had to reach Tentulbaria village by another route, which was more costly and troublesome. I travelled by a three-wheeled vehicle called Mahendro from Barguna Paurashava to Phuljhuri Ghat (river port) and then took a 20-minute boat journey instead of the two-hour boat journey. But arranging this alternative journey led me to face social pressure as my relatives from my in-laws' family advised me to use the same waterways used by the villagers.

After this journey, I understood how difficult it is for the Tentulbaria villagers to travel from their village. No one can get out of the village except by taking the boat for seven to eight months of the year. There is no surfaced road in this area. During the winter season, people can walk or men can use motorcycles on the local, muddy road, but women do not have access to any vehicle. They cannot walk such a long distance, and women of this village do not ride motorcycles. Therefore, when they become sick or pregnant, they can only use boats. However, there is no easy way to get onto the boats, as there are no stairs or any arrangement for the old, sick, children or pregnant women. The following photographs (Figures 2.2 and 2.3) show how difficult the journey in and out of Tentulbaria village is.

Figure 2.2 A vulnerable wooden bridge to reach the port to get on the boat. The bridge is half broken and has an irregular platform on it which needs extra attention to cross it.

Source: Fieldwork, 2013

Figure 2.3 Getting on the boat. It is impossible to get on the boat without the help of a strong person. This aged person in the photograph has been helped by the young man, and the woman is waiting to get the help. But normally, women do not want to take help from an unknown man, and they especially try to avoid touching. It shows how difficult it is for a woman or sick person to go out of Tentulbaria village without a man accompanying them.

Source: Fieldwork, 2013

Tentulbaria is a remote village of Barguna Sadar Upazila. There is no office, NGO centre or hotel to stay at night. I talked about accommodation with my key informant, and he arranged my accommodation at his house. It was a great help for me. During my stay, I have found that this village does not have all the telephone networks of Bangladesh. Electricity in this village is also very irregular. I was losing my mobile charges and facing communication problems with urban areas and my family in Dhaka. I have also found that due to the irregular electricity supply, not all inhabitants of this village can keep updated weather warning news during cyclones, and they have to depend on their own weather sense or that of others.

I conducted an interview with my first key informant who is a senior school teacher in Tentulbaria village. He was a very helpful informant. He and his family were directly affected by Cyclone Sidr and unfortunately lost about 33 relatives. He was also injured during that time. Other key informants were a businessman, a housewife and a student in the village. Using the snowballing method, I chose 12 respondents from different categories. There was no female businessperson, farmer or labourer found in this village, so I could not include any female respondents in those categories.

In-depth interviews with official: doctors

I selected doctors (male and female), the Mayor of Barguna municipality, the Chairman of Union Parishad and NGO officials (male and female) as my respondents.

It was challenging to get appointments with doctors. First, they did not want to be interviewed, as they thought I might be from the media or any anti-corruption offices. I introduced myself and made them understand that it was only for research and I would not have any questions on their duties and work. Following this, I got the chance to talk with a civil surgeon of the Barguna District who is in charge of all types of healthcare centres in Barguna and a female doctor who has been working in a 'Mother and child care centre' for a long time. Both of them were in Barguna during cyclones Sidr and Mahasen. Their experiences and opinions were very important and useful for my research.

In-depth interviews with official: NGOs

After Cyclone Sidr, a higher number of NGOs started working in Barguna. Some of them are working on disasters, focusing on awareness programmes. I selected these NGOs according to their ongoing programmes in Barguna: both national and local NGOs were selected if they were working on disasters, health and women. My local key respondents helped in this regard. I contacted nine NGOs, and seven of them participated in interviews, and two of them preferred meetings. They were very friendly and helpful and provided me with statistical data and information. They shared their field experiences during disasters and mentioned several cases regarding healthcare accessibility, social constraints and their limitations during and after disasters. These interviews helped me to know more precisely the limitations of the prevailing disaster management plans.

In-depth interviews with official: the Mayor and the Chairman

I talked with the Mayor of Barguna municipality and the Chairman of the Badarkhali Union to find out about their prevailing plans for disasters and the limitations and constraints they face in executing these plans. I also asked them to share their opinions on the prevailing healthcare facilities and recommendations to improve the conditions that the research identified. They also provided me two recent maps of these two study areas that are not easily available in Dhaka.

Focus group discussions and participatory diagramming

Focus group discussions and participatory diagramming methods were used to gather data and information from specific groups, because 'the hallmark of focus groups is their explicit use of group interaction to produce data and insights that would be less accessible without the interaction found in a group' (Morgan 1996, p. 2). I conducted four focus group discussions using participatory diagramming. Focus groups were formed of four to seven members in each group and were single-sex because men and women face specific health problems and difficulties around healthcare access in disasters. Group members were selected from the local inhabitants for their varying economic and education status. These methods were very helpful to gather data, especially from the women's group, because discussion within same category of members helped them to think more about the topic, and they spontaneously shared their own opinions. The venues for the discussions were chosen according to the group characteristics because 'different settings experienced under different conditions could both enhance and detract from a successful focus group experience' (Bosco and Herman 2010, p. 202). The women's group preferred houses, whereas the men's group preferred to meet at the central marketplace. Before starting the discussion, a short presentation was made about the research and its aims to make the participants familiar with the method. The topics for discussion were supplied in 'open question form', but additional points were raised by the participants. Discussions took place on prevailing healthcare facilities, factors influencing accessibility to healthcare facilities during and after disasters and the impact of gender on disasters. They also shared their opinions on what would improve conditions. I worked as the moderator in each discussion and was very careful to pay special attention to both 'problematic silences' and 'problematic speech' and create a balance between detachment and interest (Crang and Cook 2007).

The same members of the focus group discussions also participated in diagramming. Participatory diagramming aimed to reveal the personal opinion of each participant on the conditions of healthcare facilities in the area, both during normal times and in disasters. Diagrams had a scale from 1 to 5, representing the best to worst healthcare. Simple words that represent key ideas were used to make the diagramming method friendly for participants who might be illiterate, quieter or less confident (Kesby et al. 2005). The reasons behind their opinions have also been recorded during the discussions.

Figure 2.4 Old school building in Tentulbaria village
Source: Fieldwork, 2013

Observation and visits to key sites

Observation was a 'supportive or supplementary method' (Robson 2002, p. 312) in my data collection, which helped me to 'validate or corroborate the messages obtained in the interviews' (Robson 2002, p. 312). For this, I visited different places of the Barguna municipality and Tentulbaria villages to get an idea of the local area and to observe the intensity of the problems in the health and transport sectors faced by the local inhabitants during normal periods of the year. These visits showed me how vulnerable this area becomes when cyclones attack. I visited several healthcare centres in Barguna municipality, including Barguna District Hospital, the Mother and Child Health Complex, pharmacies, and educational centres and offices which become cyclones shelters during disasters. I also visited the only school of Tentulbaria village, which becomes the cyclone shelter. Visiting these centres helped my understanding of the problems reported by the respondents.

Seminar and open discussion with professionals

Returning to Dhaka, I presented my research in a seminar followed by open discussion in the Dhaka University. The seminar was arranged by the DRTMC. In that seminar, gender geographers, disaster researchers and faculties and PhD students of the Geography and Environment Department were present. It was a good opportunity for me to discuss my research and receive important and useful opinions from Bangladeshi researchers who work on gender, health and disaster in Bangladesh.

2.9 Ethical issues and considerations

Two ethical considerations arise from this present research: firstly, the implications of the research on respondents and, secondly, accessing medical records.

First, I found that most of the informants were happy to participate in the research, as this study is working on health conditions and healthcare facilities in their disaster-prone area. However, I asked for permission from every respondent before their interview and before discussing any private or sensitive issue. I also informed them that all information would be stored with full security, and the data will be used anonymously. A few of the respondents became upset talking about their experiences in disasters, at which point we took a break or changed the topic. I was prepared with an option to leave these topics out of the interview schedule if the informant wanted. I also kept with me the phone numbers and addresses of the nearest healthcare centre/healthcare facilities to use or provide the respondents if any emergency or need arose. Fortunately, there was no need to use them, and the respondents who took a break continued their discussions afterwards.

Again, I explained my research to the respondents and made them understand that this research cannot promise any development in local healthcare facilities or cyclone shelters. But I informed them significance of the research and how it would be helpful for policy makers to develop disaster management plans. I asked the respondents if they were interested to know about the final findings of the research. Most of them were not interested, except a few officials.

Secondly, in regards of accessing medical records, I only collected general medical records containing age and gender, but no personalized records. These records are accessible to any researchers who are working in the related fields, simply requiring permission from the authorities.

2.10 Insider-outsider roles in the research

Literatures on research methodology include guidance and debates over the effect of the role of the researcher as an insider or an outsider. Such considerations centre on 'the personhood of the researcher, including her or his membership status in relation to those participating in the research' as 'an essential and ever-present aspect of the investigation' (Dwyer and Buckle 2009, p. 55). In the present study, my research identities undoubtedly influenced the research process in the traditional society of Barguna. In my research, my identities were complex and multiple. Holding insider-outsider positions simultaneously led me to engage in critical scrutiny of these positions, which Mason describes as a 'reflexive act' (Mason 2002, p. 5). Reflexivity allows us to 'observe our feelings and positionality and the analysis of this dynamic becomes an important source of data' (Takeda 2012, p. 286 cited in Cui 2015, p. 357). Awareness of my insider-outsider positions thus helped me to collect data in this remote location of Bangladesh. In the following paragraphs, I explain how I occupied both insider-outsider positions in conducting my fieldwork.

I am a Bangladeshi and speak the native language of Barguna (Bangla). I know the general culture and the gender roles and responsibilities of a Bangladeshi woman well. Although I am from Dhaka, I have relatives (my sister-in-law and her family) in Barguna, and my husband is from the same administrative division of Barguna. All these factors helped to give me access as a partial insider in traditional Barguna society and helped me to conduct fieldwork in residents' private domain. Speaking the same language helped me to communicate well and to understand the respondents' opinions clearly, and this appeared to make them feel comfortable during the interviews. Being familiar with their culture and gender roles, I maintained their traditions when I visited them. I always took a chaperone with me and followed the traditional dress code, which helped me to be accepted by the local society.

Female respondents felt comfortable enough to share their very personal experiences with me, as I was a female researcher. For example, one respondent elaborately explained her experience of pregnancy during Cyclone Sidr, and many female respondents mentioned problems related to menstruation during cyclones. These are issues that many women in this region would never speak with researchers about. Above all, my insider identity was helpful for me to select the respondents through my personal network and be referred to as a daughter-in-law of a local family and thus respected and accepted. During the interviews, because respondents saw me as an insider, they shared their opinions on different problems, e.g. mismanagement and corruption in local healthcare centres and disaster management plans, which was very helpful for the research. As Miller and Glassner (2011) comment, 'researchers should be members of the groups they study, in order to have the subjective knowledge necessary to truly understand their life experiences' (Miller and Glassner 2011, p. 141).

However, beside this insider position, I occupied an outsider position, too. My home city is Dhaka, which is 319km from Barguna. This made me an outsider in Barguna. However, for the research, this had some advantages: being a Dhaka resident and an outsider allowed me to notice the differences in local culture and tradition between these two areas. I was struck by how strongly traditions are followed in Barguna. In addition, I was introduced as a researcher from the UK, which both created an outsider impression but also built some trust among the respondents, as I was introduced as a professional person. My higher educational and occupational status (doing a PhD and working as a faculty member at Dhaka University) also helped in this regard. I found that respondents were eager to answer the questions and express their opinions because they believed that my researcher position would help me to convey their messages to the right authority (government, NGOs, INGOs, policy makers etc). For example, professionals from NGOs mentioned their problems coordinating with government officials. Female respondents mentioned their experiences and opinions about gender discrimination in the family and in wider society, about which they normally do not talk. They wanted to inform wider society via my research. Meanwhile, male members expressed their helplessness in changing gendered attitudes, which became a key finding of this research.

It is clear, then, that in many ways my identity as an insider-outsider (when the two roles converged) was helpful for the research. Respondents trusted me but also took me seriously, and this encouraged them to explain their opinions. However, there were times when my insider-outsider identity was not helpful; this identity also created constraints during some interviews. For example, I selected government officials through my personal networks, but the officials took time before agreeing to appointments. I felt that my outsider identity created some suspicion amongst them. First, they enquired if my research was collecting data on corruption and mismanagement in health centres. I had to assure them that it was not and explain my research in detail before they would agree to the interview. However, I found that during the interview, they started to trust me, and my insider identity came to the fore as they opened up and mentioned the many problems they face in medical relief distribution during disasters.

Above all, these insider-outsider identities and relationships with the respondents helped me to capture data from the private domain of the traditional society in which the research was based. As Cui (2015) states, 'there is a need to pay attention to the researcher-researched relationships and its implication for research design' (Cui 2015, p. 367). She explains how she, too, had to be simultaneously an insider and an outsider in carrying out qualitative research in Chinese society and how these conflicting statuses influenced her research process. Miller and Glassner (2011) also emphasize the importance of that 'the role that social similarities and differences between the interviewers and interviewees played in producing these disparate accounts of the same phenomenon' (Miller and Glassner 2011, p. 144). Throughout my analysis, I have endeavoured to be critically aware of this possibility.

2.11 Data analysis

A grounded theory approach and 'coding' data analysis techniques were undertaken. I followed two sequential types of coding, 'initial coding' and 'focused coding'. Initial coding was undertaken by line-by-line coding, which helped me to get fresh ideas and concepts on the data, and 'focused coding', which 'uses most frequent and significant initial codes' (Charmaz and Bryant 2011, p. 303). Through constant comparing 'data with data, data with codes, codes with codes and codes with tentative categories and categories with categories' (Charmaz and Bryant 2011, p. 292), the tentative theoretical categories emerged, and I continued the process of data collection and analysis until the properties of the categories had reached saturation point. Analyzing and conceptualizing the relationships between phenomena helped me to produce 'substantive-level theory' (Robson 2002, p. 194) and the final findings of my research.

To assess the validity of my research findings, data were triangulated. Triangulation helps to compare the results from different methods of data collection and to develop or corroborate an overall interpretation (Mays and Pope 2000). Therefore, in my research triangulation of data from interviews, focus group discussions and participatory diagramming strengthens confidence in the final findings and conclusions.

2.12 Research limitations

Lack of data and information on health problems in disasters are significant problems in Bangladesh and also for this research. Even in regular times with no disasters, reliable data on vital events, including births, deaths and the incidence of diseases, are inadequate in Bangladesh (UN-HABITAT 2003), and this worsens during times of disaster. There are no data on the health impacts of disasters for specific regions. Again, available information is too narrow to give an idea of the wider context. Detailed data on types of injuries, age and male-female ratio of disaster victims are rarely available, and data on long-term health impacts of disasters and updated information on victims' conditions are not available at all.

Because of these gaps in basic data, the present study started from a research question on the impacts of disasters on health, in addition to the other questions. The methodology was planned using intensive methods to reveal in-depth data and information from the selected study area within the set time of the research programme.

Again, due to the limited time and scope of the research, I could not cover every group, such as children, older people and people with disabilities, in detail. These groups are mentioned in the research, but they deserve special focus. This research has also revealed less privileged and extremely vulnerable groups in the study area, including widows and young married women who have been left by their husbands. This group also requires specific research with a different methodology, as they are not covered in detail in the current research.

2.13 Conclusion

My research aims to reveal the gendered impacts of disasters on health condition and healthcare access. This aim includes analysis of gender and several other social factors including socio-economic conditions, culture, social attitudes and behaviour in a disaster-prone area. This topic necessitated qualitative research using in-depth interviews, focus group discussions and observation. The literature review reveals that disasters have uneven social, regional and especially gendered impacts, affecting more women than men, but still, disaster management plans are not able to focus on equity and equality for the victims in the most vulnerable, disaster-prone areas of the world. Despite several recent developments, there is still a gap in research and disaster plans which needs to be identified with the help of the victims themselves. So, my plan was to visit the disaster fields and talk with the victims in a more intimate way so that I could understand their problems, get information and collect data while being with them in their own environment. I visited the most disaster-prone region of Bangladesh to carry out my research. According to my respondents, no one visits these areas to talk to victims and find out their needs; rather, decisions are made on disaster responses from the upper levels of management authority. I was the first person who visited Tentulbaria village with the sole aim of listening to them. They were so happy and eager to share their experiences with me. One of the respondents was appreciating my long presence among them, saying, 'Many people came to visit us after two three months of cyclone Sidr . . . but they were all in such a hurry . . . no-one listened to our stories, our sufferings . . . but we like to share our experiences . . . we like to talk about it. . .' (Male, age 48, rural area). As a researcher, I was overwhelmed listening to their stories and touched by their feelings. As I analyzed the data and findings, I became more confident about the significance of this research. In the following chapters, I detail the rich findings of the research, as they allow in-depth explanation of the gendered impacts of disasters on health and healthcare access.

References

Agrawala, S., Ota, T., Ahmed, A. U., Smith, J. and Aalst, M. V. (2003). "Development and climate change in Bangladesh: Focus on coastal flooding and the Sundarbans." OECOD, France.

Ahmed, S. M., Adams, A. M., Chowdhury, A. M. R. and Bhuiya, A. U. (2000). "Gender, socioeconomic development and health-seeking behaviour in Bangladesh." *Social Science & Medicine* 51(3): 361–371.

Ahmed, S. M., Hossain, M. A., Chowdhury, A. M. R. and Bhuiya, A. U. (2011). "The health workforce crisis in Bangladesh: Shortage, inappropriate skill-mix and inequitable distribution." *Human Resource for Health* 9(3): 1–7. www.human-resources-health.com/content/9/1/3.

Ali, A. (1996). "Vulnerability of Bangladesh to climate change and sea level rise through tropical cyclones and storm surges." *Water, Air, & Soil Pollution* 92(1): 171–179.

Ali, A. (1999). "Climate change impacts and adaptation assessment in Bangladesh." *Climate Research* 12: 109–116.Baden, S., Green, C., Goetz, A. M. and Guhathakurta, M. (1994). "Background report on gender issues in Bangladesh." Institute of Development Studies, University of Sussex. IDS. Report 26.

Banglapedia. (2014). http://en.banglapedia.org/index.php?title=Cyclone, accessed on 27/6/2015.

BBS. (2006). *Population Census-2001, Community Series*. Zila, Barguna, Bangladesh Bureau of Statistics.

BBS. (2011). "Statistical yearbook of Bangladesh-2010." Bangladesh Bureau of Statistics. www.bbs.gov.bd/PageWebMenuContent.aspx?MenuKey=230, accessed on 2/5/2015.

Bern, C., Sniezek, J., Mathbor, G. M., Siddiqi, M. S., Ronsmans, C., Chowdhury, A. M. R., Choudhury, A. E., Islam, K., Bennish, M., Noji, E. and Glass, R. I. (1993). "Risk factors for mortality in the Bangladesh cyclone of 1991." *Bulletin of the World Health Organization* 71(1): 73–78.

Blaikie, N. (2009). *Designing social research,* UK, Polity Press.

Bosco, F. J. and Herman, T. (2010). "Focus Groups as Collective Research Performance." In D. Delyser, S. Herbert, S. Aitken, M. Crang & L. McDowell (eds.) *The Sage of Handbook of Qualitative Geography*. London, Sage Publication Ltd: 193–207.

Byuro, B. P. (2006). *Bangladesh Population Census, 2001: Community Series*, Bangladesh Bureau of Statistics, Planning Division, Ministry of Planning, Bangladesh, Government of the People's Republic of Bangladesh.

Cannon, T. (2002). "Gender and climate hazards in Bangladesh." *Gender & Development* 10(2): 45–50.

CDMP. (2014). *Assessment Stakeholder's Role in Preparation For and Facing the Tropical Storm Mahasen*. Bangladesh, Ministry of Disaster Management and Relief.

Charmaz, K. and Bryant, A. (2011). "Grounded Theory and Credibility." In Silverman, D. (eds.) *Qualitative Research*. 3rd edition, Thousand Oaks, CA, SAGE Publications: 291–309.

Crang, M. and Cook, I. (2007). *Doing Ethnographies*, Thousand Oaks, CA, SAGE Publications.

Cui, K. (2015). "The insider-outsider role of a Chinese researcher doing fieldwork in China: The implications of cultural context." *Qualitative Social Work* 14(3): 356–369.

DDM. (2013). *Disaster Preparedness Response and Recovery*. Bangladesh, Department of Disaster Management.

del Nino, C., Dorosh, P. A., Smith, L. C. and Roy, D. K. (2001). "The 1998 floods in Bangladesh: Disaster impact, coping strategies and response." Research Report 122, Washington, DC, International Food Policy Research Institute.

Dwyer, S. C. and Buckle, J. L. (2009). "The space between: On being an insider-outsider in qualitative research." *International Journal of Qualitative Methods* 8(1): 54–63.

Enarson, E. and Meyreles, L. (2004). "International perspectives on gender and disaster: Differences and possibilities." *International Journal of Sociology and Social Policy* 24(10/11): 49–93.

Government. (2008). *Cyclone Sidr in Bangladesh: Damage, Loss and Needs Assessment For Disaster Recovery and Reconstructions*. Bangladesh, Government of Bangladesh.

Graham, E. (2005). "Philosophies Underlying Human Geography Research." In Flowerdew, R. and Martin, D. (eds.) *Methods in Human Geography: A Guide for Students Doing Research Projects*, New York, Routledge: 6–30.

Health Bulletin 2013. (2013). Health Bulletin 2013, District: Barguna, Barguna Civil Surgeon Office, Barguna, Ministry of Health and Family Welfare, Bangladesh.

Health Bulletin 2014. (2014). Health Bulletin 2014, Upazila: Barguna Sadar, Barguna Civil Surgeon Office, Barguna, Ministry of Health and Family Welfare, Bangladesh.

Hoggart, K., Lees, L. and Davies, A. (2002). *Researching Human Geography*, London, Hodder Arnold.

Hossain, M., Islam, M. T, Sakai, T. and Ishida, M. (2008). "Impact of tropical cyclones on rural infrastructures in Bangladesh." *Agricultural Engineering International: The CIGR EJournal* 2(X). www.cigrjournal.org, accessed on 5/2/2015.

Houston, M. J. and Sudman, S. (1975). "A methodological assessment of the use of key informants." *Social Science Research* 4(2): 151–164.

Islam, A. S., Bala, S. K., Hussain, M. A., Hussain, M. A. and Rahman, M. M. (2011). "Performance of coastal structures during Cyclone Sidr." *Natural Hazards Review* 12(3): 111–116.

Islam, M. R. and Ahmad, M. (2004). *Living in the Coast: Problems, Opportunities and Challenges*, Bangladesh, PDO-ICZMP.

Kahn, R. L. and Cannell, C. F. (1957). *The Dynamics of Interviewing; Theory, Technique, and Cases*. Oxford, England, John Wiley & Sons.

Karim, M. F. and Mimura, N. (2008). "Impacts of climate change and sea-level rise on cyclonic storm surge floods in Bangladesh." *Global Environmental Change* 18(3): 490–500.

Kesby, M., Kindon, S. and Pain, R. (2005). "Participatory' Approaches and Diagramming Techniques." In Flowerdew, R and Martin, M. (eds.) *Methods in Human Geography*, London, Pearson: 144–166.

LGED. (2015). www.lged.gov.bd/ViewMap, accessed on 1/5/2015.

Loayza, N., Olaberria, E., Rigolini, J. and Christiaensen, L. (2009). "Natural disasters and Growth: Going Beyond the Averages." World Bank Policy Research Working Paper (4980).

Lowe, M. S. and Short, J. R. (1990). "Progressive human geography." *Progress in Human Geography* 14(1): 1–11.

Mahmood, S. S., Iqbal, M., Hanifi, S. M. A., Wahedi, T. and Bhuiya, A. (2010). "Are 'village doctors' in Bangladesh a curse or a blessing?" *BMC International Health and Human Rights* 10(1): 18. www.biomedcentral.com/1472-698X/10/18.

Marshall, C. and Rossman, G. B. (1999). *Designing Qualitative Research*, Thousand Oaks, CA, SAGE Publications.

Mason, J. (2002). *Qualitative Researching*, Thousand Oaks, CA, SAGE Publications.

Matin, N. and Taher, M. (2001). "The changing emphasis of disasters in Bangladesh NGOs." *Disasters* 25(3): 227–239.

Mays, N. and Pope, C. (2000). "Qualitative research in health care: Assessing quality in qualitative research." *BMJ: British Medical Journal* 320(7226): 50.

Miah, M. Y., Mannan, M. A., Quddus, K. G., Mahmud, M. A. M and Baida, T. (2004). "Salinity on cultivable land and its effects on crops." *Pakistan Journal of Biological Sciences* 7(8): 1322–1326.

Miller, J. and Glassner, B. (2011). "The 'Insider' and the 'Outside': Finding Realities in Interviews." In Silverman, D. (eds.) *Qualitative Research*. 3rd edition, Thousand Oaks, CA, SAGE Publications: 131–148.Mooley, D. (1980). "Severe cyclonic storms in the Bay of Bengal, 1877 1977." *Monthly Weather Review* 108: 1647.

Morgan, D. L. (1996). *Focus Groups as Qualitative Research*, Thousand Oaks, CA, SAGE Publications.

Network for Information, Response and Preparedness Activities on Disaster (NIRAPAD). (2007). "Cyclone SIDR kills hundreds in Barisal and Khulna division." www.nirapad.org, accessed on 15/7/2008.

Paul, A. and Rahman, M. M. (2006). "Cyclone mitigation perspectives in the Islands of Bangladesh: A case of Sandwip and Hatia Islands." *Coastal Management* 34(2): 199–215.

Prata, N., Quaiyum, M. A., Passano, P., Bell, S., Bohl, D. D., Hossain, S., Azmi, A. J. and Begum, M. (2012). "Training traditional birth attendants to use misoprostal and an absorbent devivery mat in home births." *Social Science & Medicine* 75: 2021–2027.

PreventionWeb. (2014). www.preventionweb.net/english/hazards/statistics, accessed on 19/8/2014, data source: EM-DAT: The OFDA/CRED International Disaster Database, www.emdat.be, Université catholique de Louvain, Brussels, Belgium.

Rashid, H. (1991). *Geography of Bangladesh*. Dhaka, University Press Limited.

Robson, C. (2002). *Real World Research: A Resource for Social Scientists and Practitioner-Researchers*, Blackwell, UK, Oxford.

Silverman, D. (1993). *Interpreting Qualitative Data*, Thousand Oaks, CA, SAGE Publications.

Silverman, D. (ed.) (2011). *Qualitative Research*. Thousand Oaks, CA, SAGE Publications.

Takeda, A. (2012). "Reflexivity: Unmarried Japanese male interviewing married Japanese women about international marriage." *Qualitative Research* 13(3): 285–298.

Tamima, U. (2009). "Population evacuation need assessment in cyclone affected Barguna district." *Journal of Bangladesh Institute of Planners* 2: 145–157.

UN-HABITAT (2003). *Water and sanitation in the world's cities: local action for global goals*. London, United Nations Human Settlements Programme, Earthscan publication Ltd.

Valentine, G. (2005). "Tell Me About . . . Using Interviews as Research Methodology." In Flowerdew, R. and Martin, D. (eds.) *Methods in Human Geography: A Guide for Students Doing a Research Project*, London, Pearson Education.

World Bank. (2015). http://data.worldbank.org/topic/poverty, accessed on 3/5/2015.

3 Impacts of disaster on health

3.1 Introduction

Natural disasters like cyclones involve exposure to tremendous forces of nature and can have lasting effects on an individual's mental and physical health long after the occurrence of the disaster, even for a lifetime (Sastry and Gregory 2013). Such health impacts can be direct or indirect (Goldman et al. 2014). The direct impact of a disaster causes injury, trauma, death and stress, and indirect impacts are created through the breakdown of infrastructure, reducing the healthcare provision and thereby exacerbating injury conditions, influencing chronic diseases and infections. Again, disaster outcomes are a combination or additive total of risk and resilience factors (Bonanno et al. 2010). These factors influence victims during the different periods (pre-during-after the event) and intensify health impacts of disasters. They are influenced by more than one factor while taking decisions for safe shelter before a cyclone or struggling with strong winds and storm surges and keeping themselves safe during the disaster event. Post-disaster sufferings, like going without injury treatment, damage to and losses of property, also become intensified for the victims depending on their economic ability and their position in the family or society influenced by factors like gender and economic status. These factors make the situation complicated for each person and also have varied psychological impacts on them. Even being in a safe shelter does not always enable a person to avoid panic and loss as a disaster unfolds. So, discussions of the health impacts of disasters are complex. In this chapter, the focus will be upon an intensive analysis of how inhabitants face disasters during the three phases (pre-during-after the event) to reveal a full picture of the impacts of disasters on the physical and mental health of victims. Here, the pre-disaster period is defined as the phase after the warning has been forecast until the cyclone hits the coast. The during-cyclone period is the timespan when strong winds blow and storm water surges over the land (Alam and Collins 2010). For this research, the post-cyclone period is the period after the cyclone becomes weak and the storm water starts to ebb until a normal daily routine returns to the affected areas at a basic level, i.e. the transportation system becomes usable, urban facilities resume and affected people can start their daily routines again at home.

3.2 Context: cyclone warnings and preparedness: confusion, dilemma and helplessness

The health impacts of cyclones start from the pre-disaster period when warnings are issued and people feel pressured, confused and helpless in their preparations for the cyclone attack. These include panic attacks, muscle strains, lacerations and stress-induced cardiac incidents during the cyclone preparation period (Goldman et al. 2014). Health impacts become intensified when disaster preparations are incomplete and people face the cyclone unprepared; however, some disasters exceed all expectations and can overwhelm even the most thorough of preparations (Fuse and Yokota 2012). In Bangladesh during the last decades, 'building coastal defences and cyclone shelters (CS)' (Mallick 2014, p. 655) and early warning systems have received great attention by national and local government, NGOs, disaster managers and most of the inhabitants living in the coastal region (Paul 2009; Mallick 2014). Early warnings have been one of the most popular and effective methods (CCC 2009; Cash et al. 2014; Mallick 2014), reportedly saving 'the lives of tens or hundreds of thousands'; otherwise, the death toll from Cyclone Sidr 'might have been on par with the Asian tsunami of 2004' (New York Times 2007 cited in Paul 2009, p. 290). During Sidr in 2007, weather warnings were issued on TV and the radio and by local government, NGOs and local volunteers. Respondents also received warning news updates from their relatives in other districts until electricity supplies and mobile networks failed, but confusion, a deficit of understanding about the correlation of warnings and severity of storms and a disbelief about weather warnings generally prevented many from taking shelter (Alam and Collins 2010; Islam et al. 2011; Kunreuther et al. 2009; Alam and Rahman 2014). In addition, Islam and his colleagues found in their research that non-evacuation before Sidr was also caused by a tsunami false alarm a few days before and resulted from the fact that it was a long time since the last devastating cyclone, which had been in 1991 (Islam et al. 2011). My respondents pointed out that 'when a cyclone is predicted it does not happen on that date' (Male, age 20, urban area), or 'If death is our fate going to a shelter will not save us' (Male, age 52, urban area). Many made no preparations; they were following their daily routines, such as sleeping at home on the disaster night. Some even accepted invitations away from home and left their families (even young girls and children) alone, which they acknowledged later was a 'foolish decision'. Many respondents said that if they had known that the cyclone would be so severe, they would have taken precautions and left the house before it arrived. This was also a common theme among the survivors who had been trapped in the floodwaters of Hurricane Katrina (Kunreuther et al. 2009). Most were following a 'wait-and-see' approach (Alam and Collins 2010). A coordination game (Kunreuther et al. 2009; Cash et al. 2014) was found where evacuation depended on a neighbour's decision, with people from the same neighbourhood leaving together, calling to each other when they could see the approaching storm water.

Six years after Cyclone Sidr, the situation with regard to evacuation had not changed. There was an increase in awareness but no significant increase in the number of evacuees during Cyclone Mahasen in 2013. Most people stayed at home or planned to take shelter in a neighbour's house if an emergency transpired. But these decisions – no preparation for the cyclone, taking shelter during the storm attack, staying in vulnerable houses (Paul 2009), taking risks to save livestock or valuables – created a context for physical and psychological health impacts in Barguna. Not all of the victims were physically injured, but they had to face extreme mental pressure like panic attacks, intense fear and post-traumatic stress effects and losses which have been affecting their mental health since the event.[1]

3.3 Physical health impacts of disasters: exposure to injuries and death

My findings are that neglecting cyclone warnings (Islam et al. 2011), not going to the official shelters and staying in flimsy houses were found to be behaviours associated with most casualties and injuries in Cyclone Sidr. People who went to cyclone shelters or strong buildings well before the cyclone struck did not experience any physical injuries. Other researchers have similar findings (Paul 2009; Paul 2010). Rokeya, an older woman from the rural area of Barguna, showed her annoyance at those who did not listen to the warnings and suffered during Cyclone Sidr: 'Those people who suffered and most of those who died did not listen to the warnings about Cyclone Sidr' (Female, age 65, rural area). About 11 injuries were reported among 30 of my respondents, including members of their families. Nine of the injuries were because of non-evacuation during the cyclone warning period and being unprepared. These ranged from minor cuts to life-threatening injuries. Most were due to falling trees, flying or floating objects, the strong current of the storm water or falls and slips on the muddy land while taking shelter. Some were injured saving a roof or their livestock. Similar causes of injury have been mentioned by other researchers (Schmidlin 2011; Goldman et al. 2014). Wind-related tree failure was a significant cause of death in the US between 1995 and 2007 (Schmidlin 2011), for instance, and flying debris and the collapse of all or part of a building were also significant reasons for injuries (Goldman et al. 2014).

Respondents in the present research explained in detail how they became the victims of cyclones and got injured.

Maksuda, the mother of two young girls, lives in the Barguna municipality. She and her family were not prepared for Cyclone Sidr. They did not heed the cyclone warnings and were at risk when later trying to reach safe shelter when the cyclone struck. They panicked, and her ten-year-old niece was cut on her hand by a flying object. Maksuda described the situation in her interview.

> We were sitting on sofa . . . a strong wind was blowing outside and neighbours called us to make us aware that they were leaving for safe shelter . . . but we did not hear . . . The water entered into our front yard . . . we did not notice . . . My brother came and knocked the door and in the meantime water had entered into the house and quickly it reached up to the knee level. We felt panic, we felt helpless. Now what should we do? Where should we go? There

were 7–8 persons in that house with two children . . . My husband took one daughter on his shoulder and we left the house to reach my father's house. But the wind was very strong outside . . . When I took one step, it pushed me back one step again . . . It took at least one hour to reach the house which needs five minutes normally. Trees were all toppling down . . . My niece got cut on her hand by something in that wind . . . It was only by the blessing of Allah that no tree toppled on us. . .

(Female, age 38, urban area)

Maksuda's description captures the vulnerability created for the whole family, including children, to physical injuries and mental impacts, due to lack of an in-time decision on evacuation and dependency on communication through family networks. Including messengers, all of them were exposed to the devastating attacks of Cyclone Sidr while reaching safe shelter in strong wind and storm water. This explains the reasons for most causalities and injuries happening during late evacuations. They were lucky that day, but not all late evacuees were lucky in finding shelter during the cyclone. Sahida, another respondent from the urban area, and her husband ran for safe shelter through toppling trees and strong winds. Her husband was hit by a tree on his head and was badly injured, leading to lifelong deafness.

In another case, at the 'Asrayan' (Government residential area for the landless people in Barguna municipality), two parents did not believe the warnings and left their two teenage daughters at home alone. Their house is located at the river bank side and is often flooded during high tides. During my visit to the Asrayan, I found the wrecked house, made of only tin sheets, possibly the first victim of any cyclone. The father went for his business to Dhaka, and the mother went to a relative's house to give a hand in exchange for extra money. Hearing the warnings, one of the girls became scared and went to the nearby school with neighbours before the cyclone started. The older girl, Asma, did not believe the warnings and did not leave their tin-shade (kutcha) house. In her interview, Asma described the situation she had to face during the disaster.

It was raining. I thought it was normal like other days. I closed the doors carefully and went to sleep . . . then after some time . . . I was woken by the sound of neighbours knocking . . . the water had already entered . . . I got very scared . . . and quickly the water reached up to my neck . . . It felt like death . . . I got out of the house . . . swimming and sometimes running towards the nearby school . . . Trees were breaking, toppling down everywhere . . . I got my dress (Maxi) caught in something . . . I fell . . . got hurt on the knee . . . neighbours helped me to keep going again . . . They saved my life that day. . .

(Female, age 22, urban area)

Asma's experiences shows her vulnerability to injuries and panicking after leaving the house during the disaster attack and also highlights her shocking near-death experience (Herman 1997), which she further elaborated as 'memories I would not forget for the rest of my life'. Asma's account also illustrates the problems created by ignoring the warning, by cumbersome clothing and lack of proper decisions for young people and children (discussed later) in Cyclone Sidr.

Children, older people and those with a disability, the marginalized, are the most at risk as victims of disasters due to their low status in society (Doocy et al. 2007; Rahman 2013; Marx et al. 2012; Huq-Hussain et al. 2013). Taking shelter during the disasters is very difficult for them, whereas no special arrangement or plan was found in the locality or among the family members in the study area. The situation varies for each of them according to their special needs. Children need help and supervision or to be carried to reach the shelter. In Sidr, they were taken with the others at the last moment in the strong winds and storm surge, which was a reason for their injuries and the traumatic psychological impact on them. They were found to be crying, screaming out of intense fear with their helpless parents.

While taking shelter during the disasters, parents or relatives both took responsibilities for their children, depending on their own condition, but unfortunately, sometimes not all could be saved. Kamrul from the rural area recalled how his 'baby has slipped from the lap of the mother . . . Another man rescued the baby by diving into the storm water' (Male, age 28, rural area). Rima (Female, age 28, rural area) from the rural area still mourns for her brother, whom she left to save herself while struggling in the strong storm water.

The same helpless situations were also found for older people and people with disabilities. Though knowing the location of the shelter, these people did not move from their houses because of the distance, the steep stairs into the shelter (steep stairs) or the lack of special arrangements in the toilets for them (Huq-Hussain et al. 2013). People with disabilities were found to have been left behind in vulnerable houses during the evacuation.

Staying at the vulnerable house: injuries and vulnerabilities

Sometimes, people's confidence about the resilience of their house structure, which they mention as 'quality houses' (Mallick 2014) also created confusion among them during the warning period. They depended on their houses, but their confidence drained when confronted by the force of the cyclone, and this led them to search for safe shelter at the last moment in strong winds and the storm surge. Mallick mentioned in his research that 'people tend to move towards the shelters only a few hours before the water starts to rise, by which time the wind speed may be quite high' (Mallick 2014, p. 661), and moving in strong wind and rising water was not possible for all. Runa, a respondent from a rural area, did not leave her house, which was under construction before the cyclone, and stayed with her family of 11 members with two young children. Runa described the situation in her interviews.

> We did not understand the warnings by the Red Cross. We stayed at home. At that time we were constructing our new house . . . when the water entered the house we went to the upper floor but there was no roof . . . water was everywhere . . . trees were uprooted . . . It was dangerous to move anywhere

at that time . . . We could not go anywhere . . . I had to be under the umbrella the whole night with my son of 3–4 years.

(Female, age 30, rural area)

Runa and her family were trapped in the house and remained under the open sky for whole night with wet clothes and risk of injury from flying objects.

Again, weak houses like tin-shade structures and kutcha houses (made of straw and bamboo) were partially or fully destroyed, even swept away by the strong wind and storm water, making the residents vulnerable to injury and death. Rima, from the rural area, explained in her interview that she and her family had to pay a price for believing in their house.

We thought that we did not need to go anywhere . . . All members of the family were sleeping in their beds . . . The wind was so strong outside . . . After some time the water rose . . . it blocked the doors . . . water entered with great force . . . Those asleep suddenly found themselves afloat . . . After that I do not remember exactly what happened . . . I was inside the house but found myself under the open sky . . . floating . . . whole house was destroyed. . .

(Female, age 28, rural area)

Unfortunately, Rima lost most of her family, and she herself was rescued critically injured the next morning.

Again, an older woman in Barguna was unlucky during Cyclone Mahasen in her house in the urban area. She was trapped and hurt by a fallen wall at her house, which had been hit by a huge tree. She was later rescued by neighbours and taken to the nearest cyclone shelter, but she is still facing health problems from the injury.

All these accounts explain that non-evacuation during the warning periods and taking refuge in the weak houses was not helpful for the victims but put them at great risk from the devastation of Cyclone Sidr. Their experiences emphasize the importance of taking shelter in the cyclone shelters before the cyclone hits the land, and also reveal the need for cyclone shelters in the right location and number and for other essential facilities in the disaster-prone coastal regions of Bangladesh.

Taking risks to protect valuable goods/live stock

The lack of space for cattle and other valuables, such as business stock, in the cyclone shelters and the risk of losing them to theft or burglary made several respondents reluctant to move (Alam and Collins 2010; Islam et al. 2011). They felt helpless and unprepared for a disaster. They stayed at home to guard valuable assets or took risks to transfer those things to a safe place during the disaster, which in many cases was the reason for their injuries. Hossain mentioned in her writing, 'Whereas, one of the great tragedies of Cyclone Sidr was that many of those who died knew they were at risk, yet stayed, fearful of looters' (Hossain

2008, p. 1), and she continues, 'People's solutions to crime and violence in Bangladesh are almost as detrimental as the problems themselves'.

Jasim, a young man in Barguna Municipality, had good number of cows at home, and his family did not evacuate. He described in his interviews how they tried to save their cows during the cyclone. 'There was very strong wind . . . we all went to take the cows to a safe place . . . but suddenly a tree fell on the cowshed and on the cows. We did manage to save them . . . but my uncle got hit on the head . . . by a broken branch . . .' (Male, 20, urban area).

Kamrul from Tentulbaria village described how he returned from the cyclone shelter to his exposed house to bring his cows to safety. 'I went back to my house to get the cows again, when the strong wind was blowing and trees were toppling. All of the roads were under water . . . and I took the cows swimming to a neighbour's house . . .' (Male, age 28, rural area).

Some respondents also took risks to save their houses. Alauddin described how his family members held the roof of the house from the inside to keep it flying away, but his mother and sister-in-law got cut on their hands, which could have been serious. 'We all family members . . . mother, Bhabi (wife of brother), father, brother held the roof (made of tin) from the inside of the house, to stop it flying away, which created cuts to their hands . . .' (Male, age 28, urban area).

Even, Nazem, the Muazzin of the central mosque of Barguna municipality, tried his best to save the goods of the mosque from the approaching storm water. His loneliness, helplessness, wariness and his sincerity for his duties can be seen in the following quotation. 'I felt very lonely, panicky and helpless when I saw the approaching storm water on the main road. It was dark and there was no electricity in the whole bazaar. I was alone . . . I started collecting all the white clothes and carpets from the floor of the mosque, they were very heavy. I was trying hard, I saw water entering the roads . . . suddenly a tree fell on our room . . . then I got really scared . . . but I kept continuing . . . I managed to save everything' (Male, age 28, urban area). Nazem did not leave or go upstairs in the mosque to take shelter but rather completed his duty of saving the goods of the mosque.

All these experiences demonstrate the reasons and mode of decisions taken by the victims to protect their valuable goods, increasing their vulnerability during disasters. Lack of space for cattle and goods in the cyclone shelters, as well as the importance of these valuable goods in their lives, should be considered by the disaster management authority before planning for constructing cyclone shelters.

3.4 Post-disaster periods: lack of basic needs, injuries and diseases

Impacts of the disaster on health did not finish with the end of the cyclone but rather intensified in the post-disaster period (CCC 2009). According to most respondents, 'more suffering and problems were created during the post-disaster period than during the cyclone itself'. Jasim, a male respondent from near the Barguna municipality explained the situation in the few days after Sidr. 'Sufferings are higher after disasters . . . all the crops were destroyed, the trees were all broken, the roads were blocked, nothing could come from the outside . . . the people were starving, the tube wells were under water and there was a lack of fresh water . . . Being dependent on dirty water people became sick' (Male, age 20, urban area). Indirect health impacts were observed by the survivors during the post-disaster period. During the first 24 hours after the cyclone, they faced a lack of emergency services and treatments for injuries, diseases and sickness. There was a scarcity of food, safe drinking water, and sanitation, and houses were damaged and transport systems disrupted (CCC 2009; Paul et al. 2012). Relief aid and medical help did not reach all of the affected areas, especially the remote locations, for 2–3 days after the cyclone (Islam et al. 2011). This unavailability brought long-term health problems for the victims (elaborately discussed in the next chapter). No cooked food reached Tentulbaria village until day 2–3 days after Sidr, except for some dried foods which were low in nutrition (Government 2008; Islam et al. 2011). In the Barguna municipality and surrounding areas, cooked food was distributed by the local authority in some areas but most victims were without proper nourishment for days.

Post-disaster sufferings were intense among the children, older and disabled people. They lacked food and drinking water. There was no special arrangement for them in the cyclone shelter. Children cried the whole night and the day after Sidr. Faruk explained the situation created in Tentulbaria village at the school building where older people and children were taking shelter. He also described his initiative to mitigate their suffering.

> At the Fazar time [early morning before sunrise] an old person started feeling sick for lack of food . . . I tried to get the food . . . I swam to my house . . . but nothing could be recognized . . . Still I did eventually reach my house . . . I brought some naru [dried sweets] which were kept near the roof ceiling and some papayas . . . I gave all the naru to the elders and papaya to the children . . . dividing them into pieces
>
> (Male, age 48, rural area).

Like Faruk, local inhabitants in most affected areas tried their best to help each other to survive the difficult time of the post-disaster periods and showed their solidarity with the victims (Alam and Collins 2010). People received help from their relatives and community members (Cash et al. 2014). As Jasim explained in his interview, 'We the local people helped the victims for first two days . . . after that relief began to arrive gradually' (male, age 20, urban area). Beside food, local inhabitants shared and donated clothes to the victims. Maksuda and her family survived the two days after Cyclone Sidr on food supplied by her relatives. As she says, 'My sister and my mother sent me cooked food for the next days of Cyclone Sidr as I lost my cooking arrangements' (Female, age 38, urban area). Without help from neighbours, relatives and local people, it would have been very difficult for the victims to survive during this post-disaster period. Similar findings are published from other research with Bangladeshi cyclone survivors (Alam and Collins 2010; Paul and Routray 2011; Cash et al. 2014), Hurricane Katrina survivors in 2005 and Asian Tsunami survivors in 2004 (Jacob et al. 2008). Paul and

Routrary found that 'under extreme situations, when such disasters surpass the shock-bearing capacity of the victims, informal risk sharing mechanisms through social bonding and social safety-nets become vital for short-term survival and long-term livelihood security' (Paul and Routray 2011, p. 477).

Several respondents mentioned in their interviews that due to lack of safe drinking water, safe food and adequate sanitation, different water-borne communicable diseases as stomach upset, diarrhoea, and dysentery increased among the cyclone victims. Most of the drinking water sources were damaged (tube wells) or contaminated (open ponds) with carcasses of dead bodies, animals or fish and mixed with debris and raw sewage (Paul et al. 2012). The electricity and water supplies were off for a month in the Barguna municipality, with the attendant risk of epidemics. Survivors had to collect drinking water from far off. Fortunately their awareness about the safe water and a supply of water purification tablets helped to prevent any epidemics (Paul et al. 2012). But other diseases like cold, fever, allergy, pox, blisters and especially skin diseases were spread among the inhabitants of the flooded areas after the cyclone Sidr because many were standing or swimming in dirty water for the whole disaster night. Especially in urban flood affected areas, respondents reported that water from the waste water drains got mixed with storm water and became highly contaminated. Lack of clean water and unsanitary conditions intensified these health problems among Cyclone Sidr victims, as they did also for Hurricane Katrina survivors. Scabies/lice infestation and carbon monoxide poisoning were reported among the shelter occupants after Hurricane Katrina (Jacob et al. 2008).

However, victims of Sidr did not receive proper treatments for these diseases. They depended on distributed oral saline for the diseases like stomach upset, diarrhoea, and dysentery. But treatment for the other diseases and sickness was delayed as they waited until the damaged transport system became accessible, which exacerbated the health condition of the victims.

3.5 Psychological impacts of disasters

'Disasters are traumatic events that may result in a wide range of mental and physical health consequences' (Neria et al. 2008. p. 467) and 'at the moment of trauma, the victim is rendered helpless by the overwhelming force' (Herman 1997, p. 33), which may lead to two traumas for the victims (Kleber et al. 1995). First, the occurrences of the traumatic event, cyclone itself and, second, injuries, loss of family members, damage and the destruction of community life and social contacts.

As mentioned before, in the present research, respondents mentioned two cyclones as the major disasters in their lives: Sidr in 2007 and Mahasen in 2013. Among those, Cyclone Sidr was mentioned as the deadliest, when victims became scared and panicked for the security of life, worried and anxious for the family members and upset, helpless, devastated, and frustrated for the loss, injuries and damage during the pre-during-post-disaster periods. Some of their descriptions have similarities to the syndromes of post-traumatic stress disorder : hyperarousal, intrusions, constriction (Herman 1997) and other mental consequences of traumatic events. The findings of other published research also show the existence of the psychological impacts of Cyclone Sidr on victims of other areas. A survey was conducted by the Bangladesh Association of Psychiatrists in 2007, having assessed 750 survivors just two months after Sidr, and they found that about 25.2% had post-traumatic stress disorder, 17.9% had a major depressive disorder, 16.3% had somatoform disorder, and 14.6% had a mixed anxiety and depressive disorder (Turpin 2011 cited in Nahar el al. 2014). But there is not much evidence-based data in Bangladesh because mental health impacts have not yet received enough attention (Cash et al. 2014; Nahar et al. 2014) and unlike physical injuries, mental health outcomes of disasters may not be immediately apparent; they require a systematic approach to identification (North and Pfefferbaum 2013). Mental health assessments are mostly conducted in high-income countries and evidence from low-and middle income countries is rare (Phalkey et al. 2012 cited in Marx et al. 2012). No survey was conducted to reveal any problems in the present study area and no victims were consulted about their issues by medical staff, leaving psychiatric problems unidentified.

Years after Cyclone Sidr some respondents mentioned that they had recovered psychologically and became confident about their capability against cyclone attack, which was attested by Mahasen in 2013. Bonnano et al. (2010) mentioned in his research that 'although the stakes are often higher with disasters, the same basic patterns of distribution of dysfunction and resilience have been observed' (Bonanno et al. 2010, p. 3). My respondents mentioned that having survived the mighty Cyclone Sidr, they feel capable of fighting back against future cyclones. They have also become very aware about warnings, taking shelter in a timely fashion and making plans to mitigate disasters. Again, some respondents expressed both feelings about cyclones: fear but also the strength to fight back.

Below, I report the evidence I have collected that support these findings and explains the context, reasons and consequences of the mental health impacts of cyclones in Barguna.

Psychological traumas and panics: reasons and consequences

During the interviews, respondents talked about their experiences and shared their feelings of the Cyclone Sidr and Cyclone Mahasen but due to its severity, Cyclone Sidr became the major focus of discussions. The traumatic impacts of Sidr started from the warning period with panic attacks, feelings of intense fear and helplessness. Victims became confused and scared at the sudden onset of the cyclone. They sought religious help (Alam and Collins 2010) and even got ready to embrace death. Rokeya, an aged female victim stated that she did 'Touba', the last rites for a dying person. She described her response in her interview, 'I was highly scared in Sidr . . . I did the Touba in the evening, just before cyclone attacked us' (Female, age 65, rural area). Victims got terrified when they saw that the strong wind and storm surge of Cyclone Sidr was approaching their locality with huge force. Respondents also mentioned their intense fear. Sifa described her experience in the interviews: 'I was so upset and scared that I screamed, cried whole night . . . I was afraid seeing the huge waves which filled the whole area and turned it into the sea . . .' (Female, age 50, rural area).

But not all victims expressed fear; some felt numb or stopped running when racing against the approaching storm water and could not even remember their prayers, the primary defence for the religious-minded society of Barguna. According to Herman, 'when a person is completely powerless, and any form of resistance is futile, she may go into a state of surrender. The self-defence shuts down entirely' (Herman 1997, p. 42). Rokeya stopped in the middle of her race to take safe shelter, when she was chased by huge waves of storm water and Rima forgot her prayer to the Creator: 'I was scared, helpless and panicky while struggling in the water . . . I forgot all the Suras [prayers] . . . I tried but could not' (Female, age 28, rural area). Rima and most respondents of this present research belong to a society with strong religious beliefs. Depending on the Creator and prayer is one of the first actions taken by these victims when any calamity falls on them (Alam and Collins 2010). Another respondent, Liza, could not move from her sitting position for hours until the cyclone weakened.

Victims also disregarded the pain of injuries while facing the cyclone. Sifa felt nothing when she cut her leg: 'I was still so distressed the day after Cyclone Sidr that I did not notice that I cut my leg while returning to my house . . . When I arrived I saw lot of bloods on my foot . . .' (Female, age 48, rural area).

Again, Samsun, a mother, was unable to care for her children during the disaster night when she 'froze'. She forgot her parental responsibilities: 'I was really scared, I cried and waited sitting in one place the whole night . . . holding my newborn baby . . . but I could not take care of my other children . . .' (Female, age 48, urban area). But this situation of a mother became a reason of vulnerability among their children. Panic was noticeable among the children, older people and the disabled during the pre-, during-, and post-disaster periods (Jacob et al. 2008). Their dependency for taking shelter or being rescued, and their special needs for food, clothing and medicines created trauma for them. Children were mostly taken care of by their parents, but the anxiety and vulnerability of the parents and the contagious traumatic atmosphere during the cyclone attack had a great influence on the children (Silverman and La Greca 2002 cited in Terranova 2009), 'who are powerless in comparison to adults' (Herman 1997, p. 60). The children became the most vulnerable victims, especially when their elders did not have a proper plan for a safe shelter before the disaster attack. Seeing parents crying or getting nervous or lost during the disaster attack, most of the children became terrified. The following quotations highlight some of the emotions.

Papri: 'I fainted . . . We were all safe . . . but seeing me senseless . . . my three children started crying . . . Another man came forward hearing their cries and helped us' (Female, age 38, urban area). Papri's description shows her child's mental condition after seeing her mother fainting while taking shelter.

Maksuda described the situation in a sad voice:

> Hearing people screaming 'water' . . .'water' . . . we ran towards the road (high street) . . . there were people everywhere, all of them panicking . . . My daughter of 5 years was crying and asking 'Ma what will I do, you all can

swim away in the storm water but what will happen to me?' . . . I could not hold back my tears

(Female, 38, urban area).

All these quotations identify the vulnerability of children due to lack of proper precautions by their parents. Both adults and children became exposed to the psychological impacts of disasters; intense fear, panic and in some cases long-term impacts like post-traumatic stress disorder.

Again, staying at home did not help respondents feel secure; rather they were terrified for several reasons. During the peak hours of the cyclone, respondents were in continuous tension thinking of the strength of their houses against the increasing force of Sidr, and some of their houses were damaged while they were inside, making the situation worse.

The cyclone shelters offered physical protection but staying there was a source of psychological trauma for many victims. 'Trauma clearly has a contagious effect' (Figley and Kleber 1995, p. 84) and victims as well as witnesses are highly influenced by traumatic events, many became nervous, stressed and they panicked when observing other injured or traumatized survivors. They mentioned in their interviews that memories of cyclone shelters became unforgettable for their whole life. Asma from the Asrayan, described 'It is impossible to describe the miseries of that day, everybody was crying, screaming when the water was attacking the area with high force' (Female, age 22, urban area). These situations worked as 'contagious' trauma among victims of all ages. The situation was extreme at the cyclone shelter in Tentulbaria village. As an old school building it was itself vulnerable and unfit for a large number of people. It reportedly trembled with the heavy weight and force of the cyclone. Rural respondents mentioned that if Sidr had continued more than half an hour the building would have collapsed. All the people there were crying and praying for the whole night because they felt insecure.

The separation from families, from pets, valuable assets and familiar environments also influenced the mental health condition of the evacuees. Liza, Sifa, Maksuda, Kamrul and others mentioned their burden of stress after reaching the cyclone shelters. According to the literature, 'the greater the loss of the familiar social and physical environment, the greater is the adverse impact on mental health and social adjustment' (Jacob et al. 2008, p. 564). But due to insufficient or the absence of psychological needs assessment among Cyclone Sidr evacuees, the mental health impact of disasters in Barguna cannot be revealed. By comparison a psychological needs assessment of Katrina evacuees in Houston shelters revealed post-traumatic stress disorder symptoms among the evacuees from moderate to severe among 39% and 24% of evacuees respectively (Coker et al. 2006 cited in Jacob et al. 2008).

Post-disaster period: period of mourning

The post-disaster period becomes one of frustration and grief for those who lost family members (Bonanno et al. 2010). In a very sad voice, Rima from the rural area described the situation of day after Sidr. She was rescued senseless, and when she got her consciousness back, she found that most of her family members were lost. 'I got my senses back when I heard my brother-in-law was crying, saying . . . "I have lost everyone . . . all of them are dead" . . . everybody was crying there. . .' (Female, age 28, rural area).

Victims became worried for their family members who lived far away. Jacob said, 'Indeed, separation from loved ones and familiars is generally a greater stressor than physical danger itself' (Jacob et al. 2008, p. 563). So, survivors took a risk to travel through the disrupted transport to collect the news of their nearest and dearest (Alam and Collins 2010). Nazem went home from Barguna municipality the afternoon after Cyclone Sidr to his village home to get the news of his family, and he experienced the devastation first-hand.

> I was worried about my mother . . . I saw dead bodies on the roads . . . it was scary . . . dead bodies were under the trees, floating on the water, hanging on the bamboos . . . I was so scared, I was so worried about my family . . . what has happened to them? . . . I kept walking . . . I reached home . . . by the blessing of Allah they were all fine . . . my mother started crying when she saw me.
> (Male, age 28, urban area)

The post-disaster situation was worst in Tentulbaria village. Dead bodies were all over the village. The morning after Sidr, the victims realized its severity after returning to their homes from the cyclone shelters. Rural respondent Faruk mentioned in his interview that 'there were 17–20 dead bodies stuck at my front door . . . and all of my chickens were dead in their coop' (Male, age 54, rural area).

The victims were also burdened with injuries and sickness without proper treatment. Besides, injured victims have a greater likelihood of developing post-traumatic stress disorder than other groups (Galea et al. 2005). The post-disaster period made some victims feel ambivalent about their survival. Paru, for instance, said, 'I wish I had died in the cyclone rather than facing these sufferings' (Female, age 30, urban area). Helpless in surviving the difficulties of post-disaster periods and observing the deaths and loss of the family members and neighbours and especially, total devastation of their known environment, made victims frustrated, upset, lost, and bewildered during the post-disaster periods.

Long-term psychological impacts of cyclones and resilience

Cyclone Sidr in 2007 left mixed and complicated feelings amongst the Barguna victims. Nowadays, most of them feel terrified when a cyclone warning is issued. Most people become very scared thinking about a repetition of Sidr. They feel insecure for themselves and their children. Badar mentioned his feelings: 'When higher signals are forecast . . . I start feeling scared until the cyclone hits . . . I am afraid of water . . . I become worried about the security of my children . . . and if water enters where I will keep them safe . . .' (Male, age 45, urban area). Samsun, a Sidr victim, mentioned her fear of cyclone warnings, 'Experiences of Sidr were so scary that I still feel terrified at cyclone warnings . . . I become worried thinking of its scariness . . .' (Female, age 48, rural area). Like Samsun, other respondents also reported that miserable memories and losses in Cyclone Sidr created insecurity and confusion about self-capacity and made them panic during the cyclone warnings. Some of Sidr's victims felt more than general fear. Here it should be noted that not one of the respondents visited any medical person or institute to identify their mental health problems; but their descriptions have similarities to the symptoms of post-traumatic stress disorder, and they suffered in this way for all these years after Cyclone Sidr. While we cannot formally diagnose them, we can observe that their descriptions sound very similar to hyperarousal, intrusions such as flashback recollections of the disaster and a sense of constriction due to fear, like 'freezing' (Herman 1997) during the cyclone warnings. These respondents mentioned that sometimes they feel these feelings during normal times of the year, but that any warning brought extreme mental stress associated with several physical reactions to them, which increased their vulnerability. Mahasen cyclone warnings became a trauma trigger for them. Evidence collected from the interviews is described in the following paragraphs to bring focus to the mental health problems among the Cyclone Sidr victims.

Kamal, a respondent from the Tentulbaria village, lost 33 relatives from his wider family and was injured himself during Cyclone Sidr. That cyclone has left extreme traumatic memories for him. He feels low self-esteem about surviving in cyclones. He feels terror of Sidr and every cyclone warning; even thinking of cyclone warnings brings intense fear for him. He feels physical reactions to his stress. 'When I hear warnings I feel panic, my mouth becomes dry, it feels like I am going to lose something . . . I feel insecure . . . I become scared and worried if I am going to be alive . . .' (Male, age 48, rural area). Here it should be mentioned that Kamal described himself as a normal person, maintaining all his family duties, but sometimes, memories of Cyclone Sidr and cyclone warnings make him feel different, sad and even depressed. But he did not seek any medical help.

Liza, another respondent from the rural area, suffers from the experiences of Cyclone Sidr as flashbacks during the waking state (Herman 1997). Liza mentioned in her interview that 'I still remember . . . it just appears in front of my eyes . . . the dead man who tried to save himself by lashing himself to a tree but died anyway'. She feels horrified about cyclones and mentioned her experiences during Cyclone Mahasen: 'During Mahasen, I remained stock still for 2–3 hours until the cyclone moved away . . . I became so scared that my hands and legs became stuck . . .' (Female, age 35, rural area).

Their memories of Cyclone Sidr also prompt a similar state of 'hyperarousal'. Papri (Female, age 38, urban area), a cyclone victim, fainted while taking shelter after a rumour of storm surges after Cyclone Sidr. She described her mental situation elaborately in her interview. This interview is quoted as follows.

Papri: Seven days after Cyclone Sidr . . . in the evening we heard a rumour that water would again attack us . . . I was lying with my children . . . but suddenly we had to run for our lives as people were screaming 'water', 'water' . . . I started running towards the nearest multi-storey building through an unused road, full of garbage . . . We were all bare foot . . . I thought if water came we would all die . . . I was worried for my son who was with his aunt . . . I ran as best I could . . . I climbed up and up . . . even when I was on the top floor (4th Floor) of that building, I did not feel satisfied . . . Then I fainted . . . It was only a rumour . . . Later, they [my family] brought me back home but I was not in my right senses for the whole night.

Researcher: Did you seek medical help?

Papri: No.

Researcher: Did you face any problems later?

Papri: No, it did not happen again . . . Only on that day.

Papri did not face any further health problems like this, but she now feels differently about cyclone warnings. Every warning becomes a 'disaster' for her: 'I feel terrified if a cyclone is forecast . . . every warning is a "disaster" for me . . . We might die this time . . . Last time (Cyclone Sidr) we were lucky, Allah helped us' (Female, age 38, urban area).

An extreme case of a stress-induced cardiac incident was reported during the interviews. Respondents mentioned that 'being in a state of panic', a man had died in Barguna during the warning period of Cyclone Mahasen.

All these accounts show the evidence of long term mental health impacts of disasters on cyclone victims, but due to lack of assessment, these psychiatric problems remain unidentified. According to Kamal, Liza, Alauddin and Papri, they never felt the same after Cyclone Sidr. They feel insecure, frustrated and even depressed. Memories of Cyclone Sidr haunt them even in the normal time of the year and increase during the cyclone warning time.

Similarly, besides the adults, children face negative mental impacts of disasters (Catani et al. 2008; Goldman et al. 2014), but due to the lack of any survey, the psychological problems of children after Cyclone Sidr have not been properly identified in Barguna, as they were among the tsunami-affected children in southern Thailand by Thienkrua et al. (2006). Their results show that children who experienced extreme panic or fear had a nine times higher risk of posttraumatic stress disorder symptoms compared with children without those experiences. Again, children who had felt their own or a family member's life to have been in danger had a six times higher risk of depressive symptoms. But here it should be mentioned that respondents did not report any continuing psychological problems among the children at the time of the interviews. One reason might be that the selection of respondent's age in the present research was over 18 years, and children were not intensively studied. Another reason might be 'despite their vulnerability, children typically exhibit a natural resilience in the aftermath of extreme adversity' (Masten 2001 cited in Bonanno et al. 2010, p. 15), and many of those who experience acute disaster-related distress recover within the first year after the event (La Greca et al. 1996 and Weems et al. 2010 cited in Bonanno et al. 2010). A study of children exposed to Hurricane Andrew in 1992 showed that 39% of children who met a probable post-traumatic stress disorder criterion with the first three months of the disaster declined to 18% within ten months of the disaster (Green et al. 1994; La Greca et al. 1996), and, over longer periods of time, almost all of them return to baseline levels of adjustment (Green et al. 1994 cited in Bonanno et al. 2010, p. 15).

Recovery and resilience

'Disasters produce multiple patterns of outcome, including psychological resilience', and 'some survivors recover their psychological equilibrium within a period ranging from several months to 1 or 2 years' (Bonanno et al. 2010, p. 1). Recovery and increase in mental strength were also noticeable among the respondents after Cyclone Sidr. These positive psychological outcomes are sometimes refer to as 'Post-Traumatic Growth' (PTG) (Curtis 2010, p. 75). Some of these victims mentioned that the experience of a traumatic event had helped to increase their confidence about their own abilities. After the cyclone, they became stronger and better prepared with plans to mitigate any forthcoming disaster. As time passed, some of them mentioned that they had achieved a tolerance capacity for traumatic experiences. Now they consider cyclones as natural incidents, and they have to fight back to survive. A similar coping strategy was also noticed among Tsunami-affected victims in India (Rajkumar et al. 2008). Rajkumar et al. mentioned in their research that 'in spite of incomplete reconstruction of their lives, participants reconstructed meaning for the causes and aftermath of the disaster in their cultural idiom' (Rajkumar et al. 2008, p. 846).

Among respondents in the present research, Mintu mentioned that 'We became stronger having experienced several disasters . . . five or six signals might be scary for you but we do not bother about these signals . . . We are mentally strong now . . . We are fearless' (Male, age 52, urban area).

Now, most of the respondents make preparations during the warning period and become mentally organized to face the event. They try to find a safe shelter in good time with their family members. They feel that the experience of Sidr has enabled them to think positively about the cyclones. Sardar:

> If a warning is issued we make preparations as we know a cyclone may come . . . I do not get scared . . . I think nothing more severe than Sidr could happen, besides I can swim and so I will fight back against the storm water. . .
> (Male, age 28, rural area)

Other respondents became more confident about their own abilities. Some can swim, and all can recognize the signs of danger. They also rely on their family members and religious beliefs. Together, these feelings helped the inhabitants to avoid panicking during Cyclone Mahasen.

Maksuda: No, I do not get scared now during the warning periods . . . I am much more aware. I know water will come from the river . . . I keep watching the water levels of the river (Female, 38, urban area).

Also, as time passed, some respondents felt that the intensity of fear and anxiety had faded, and they became stronger and more resilient.

Faruk: Now I can talk about Cyclone Sidr . . . The first two years it was impossible to discuss it without tears . . . Now I feel inspired to talk about my Sidr experiences . . . and people also want to listen' (Male, age 48, rural area).

Traumatic cyclones, risk and resilience factors

'With severe enough traumatic exposure, no person is immune' (Herman 1997, p. 57), and people do not react to disasters in the same way (Bonanno et al. 2010). But the burden of the mental health impacts varies with each person, culture and society (Kleber et al. 1995, p. 1), following socio-demographic and background factors, event exposure characteristics, social support factors and personality traits (Neria et al. 2008). Irrespective of gender and socio-economic conditions, the impact of traumatic events depends to some degree on the resilience of the affected person (Herman 1997) and their personality traits (Neria et al. 2008). In the present research, victims with strong self-confidence did the best during Cyclone Sidr regardless of their gendered status. Some of them fought back against the strong wind and water and saved themselves and others, regardless of being pregnant or injured. Some people were also found to be more confident against future cyclones, depending on their own knowledge and skills such as swimming. Similar findings were also evident in Bonanano et al.'s (2010) research, and they mentioned that 'hardy individuals tend to believe that they can control or influence the outcome of events and tend to reframe or reconceptualise stressful life events as challenges rather than threats' (Bonanano et al. 2010, p. 19). It is worth mentioning that Bonanno and colleagues also revealed that although many people are psychologically harmed by disaster, a great many others manage to tolerate the consequences with minimal psychological cost.

There are also a few 'resourceful individuals' (Herman 1997, p. 60) found among the victims of Cyclone Sidr. They have excellent mental strength and follow their daily routine during disasters. They went to sleep in partially damaged houses on the night of Cyclone Sidr. Both males and females in this group showed remarkable mental strength. The following two quotations elaborate this point.

Maksuda: We came back to our house after 1 o'clock . . . It was impossible to recognize my house . . . All of the trees were broken or fallen . . . We entered into the house crawling through the neighbour's house . . . The storm water had left the house covered in mud . . . After some time we went to bed . . . the next day (Female, age 38, urban area).

Jasim: One of our neighbour's trees fell on our house . . . We weren't hurt, only my uncle . . . He was injured while taking the cow to a safe shelter . . . We treated him with pan pata (betel leaf) and Detol (antiseptic solution) . . . Then we went to bed . . . The wind was still blowing strongly at that time (Male, age 20, urban area).

Jacob et al. 2008 also revealed that such people after Hurricane Katrina, even though directly affected, were able to 'remain calm and behave in an orderly and considerate fashion' (Jacob et al. 2008, p. 562).

Besides the personal traits, in the present research, socially determined factors such as gender and socio-economic conditions were found to be the most important in influencing vulnerability to mental health impacts of cyclone disasters. Similar factors among the cyclone victims have also been reported in other research by Nahar et al. in 2014. Social factors like gender and socio-economic conditions will be elaborately discussed in the following sections.

3.6 Conclusion

This chapter has discussed the physical and psychological health impacts of disasters. Drawing on rich empirical evidence from respondents, the chapter revealed the complexity of health problems created during and after disasters, describing the contexts, reasons and consequences of unequal vulnerability among the victims. First of all, the chapter highlighted that if preparation for a safe shelter is not taken during the warning periods, the highly vulnerable will need evacuation or have to stay in weak structures during the cyclone attack. These situations create intense physical and psychological health problems. Secondly, staying in cyclone shelters and strong houses reduces the risk of physical injury, but victims nevertheless were found to face psychological impacts. Thirdly, these health problems may turn into long-term health problems depending on the difficulties created during the post-disaster periods.

This chapter has also revealed that the demographic factor, e.g. age, has a direct influence on the experiences of disaster victims. Children and older people become highly vulnerable to the physical and mental health impacts of cyclones, especially if evacuation is not conducted during the warning period.

Several factors – socio-cultural, health and environmental factors – and their impacts on health are described and their significance evaluated separately in Chapter 4.

Note

1 Barguna has also experienced two other cyclone warnings, for Cyclone Nargis in 2008 and Cyclone Aila in 2009, but these both changed their track and did not reach the district (Casey 2008 and Mehedi et al. 2010). During the interviews, in response to the question on the disasters of the last ten years, most respondents mentioned the name of cyclones Sidr, Nargis, Aila and Mahasen, the most severe being Sidr and Mahasen. Experiences of Cyclone Sidr were elaborately mentioned in their interviews and came up in every section of the discussions, whereas due to minor damages, Mahasen was not highlighted, especially in the post-disaster period. Several respondents also talked about a sudden false warning or rumour of a high surge seven to eight days after Sidr, which created a great panic among the inhabitants.

References

Alam, E. and Collins, A. E. (2010). "Cyclone disaster vulnerability and response experiences in coastal Bangladesh." *Disasters* 34(4): 931–954.

Alam, K. and Rahman, M. H. (2014). "Women in natural disasters: A case study from southern coastal region of Bangladesh." *International Journal of Disaster Risk Reduction* 8: 68–82.

Bonanno, G. A., Brewin, C. R., Kaniasty, K. and Greca, A. M. L. (2010). "Weighing the costs of disaster consequences, risks, and resilience in individuals, families, and communities." *Psychological Science in the Public Interest* 11(1): 1–49.

Casey, M. (2008). "Before cyclone hit, Burmese delta was stripped of defences." *International Herald Tribune*, 9 May. www.nytimes.com.

Cash, R. A., Halder, S. R., Husain, M., Islam, M. S., Mallick, F. H., May, M. A., Rahman, M. and Rahman, M. A. (2014). "Reducing the health effect of natural hazards in Bangladesh." *The Lancet* 382(9910): 2094–2103.

Catani, C., Jacob, N., Schauer, E., Kohila, M. and Neuner, F. (2008). "Family violence, war, and natural disasters: A study of the effect of extreme stress on children's mental health in Sri Lanka." *BMC Psychiatry* 8(1): 33.

CCC. (2009). *Climate Change, Gender and Vulnerable Groups in Bangladesh*. Dhaka, Climate Chamge Cell, DoE, MoEF, Component 4b, CDMP, MoFDM: 1–82.

Coker, A. L., Hanks, J. S., Eggleston, K. S., Risser, J., Tee, P. G., Chronister, K. J., Troisi, C. L., Arafat, R. and Franzini, L. (2006). "Social and mental health needs assessment of Katrina evacuees." *Disaster Management & Response* 4(3): 88–94.

Curtis, S. (2010). *Space, Place and Mental Health*, UK, Ashgate Publishing, Ltd.

Doocy, S., Rofi, A., Moodie, C., Spring, E., Bradley, S., Burnham, G. and Robinson, C. (2007). "Tsunami mortality in Aceh province, Indonesia." *Bulletin of the World Health Organization* 85(4): 273–278.

Figley, C. R. and Kleber, R. J. (1995). "Beyond the 'Victim': Secondary Traumatic Stress." In Kleber, R. J., Figley, C. R. and Gersons, B. P. R. (eds.) *Beyond Trauma: Cultural and Societal Dynamics*. New York, Plenum Press: 75–95.

Fuse, A. and Yokota, H. (2012). "Lessons learned from the Japan earthquake and tsunami, 2011." *Journal of Nippon Medical School* 79(4): 312–315.

Galea, S., Nandi, A. and Vlahov, D. (2005). "The epidemiology of post-traumatic stress disorder after disasters." *Epidemiologic Reviews* 27(1): 78–91.

Goldman, A., Eggen, B. and Murray, V. (2014). "The health impacts of windstorms: A systematic literature review." *Public Health* 128(1): 3–28.

Government (2008). Cyclone Sidr in Bangladesh: Damage, Loss and Needs Assessment For Disaster Recovery and Reconstructions. Ministry of Finance, Government of Bangladesh, Bangladesh.

Green, B. L., Grace, M. C., Marshal, G. V., Kramer, T. L., Goldine, C. G. and Leonard, A. C. (1994). "Children of disaster in the second decade: A 17-year follow-up of Buffalo Creek survivors." *Journal of the American Academy of Child & Adolescent Psychiatry* 33(1): 71–79.

Herman, J. (1997). *Trauma and Recovery: The Aftermath of Violence – From Domestic Abuse to Political Terror*. New York, Basic Books.

Hossain, N. (2008). "The price we pay." *FORUM, a Monthly Publication of The Daily Star*, 3(1). www.the dailystar.net/forum/2008/january/price.htm, accessed on 22/9/2014.

Huq-Hussain, S., Islam, M. S. and Habiba, U. (2013). "Documentation of coping strategies of coastal people particularly the Persons with Disabilities (PWDs) living in Bagerhat District." Final Report submitted to Action on Disability and Development (ADD) International, Disaster Research Training and Management Centre (DRTMC), Dhaka University.

Islam, A. S., Bala, S. K., Hussain, M. A., Hussain, M. A. and Rahman, M. M. (2011). "Performance of coastal structures during Cyclone Sidr." *Natural Hazards Review* 12(3): 111–116.

Jacob, B., Mawson, A. R., Payton, M. and Guignard, J. C. (2008). "Disaster mythology and fact: Hurricane Katrina and social attachment." *Public Health Reports* 123(5): 555–566.

Kleber, R. J., Figley, C. R. and Gersons, B. P. R. (eds.) (1995). *Beyond Trauma: Cultural and Societal Dynamics*. New York, Plenum Press: 1–9.

Kunreuther, H., Meyer, R. and Michel-Kerjan, E. (2009). *Overcoming Decision Biases to Reduce Losses From Natural Catastrophes*, USA, Risk Management and Decision Processes Center, The Wharton School of the University of Pennsylvania: 1–27.

La Greca, A. M., Silverman, W. K., Vernberg, E. M. and Prinstein, M. J. (1996). "Symptoms of posttraumatic stress in children after Hurricane Andrew: A prospective study." *Journal of Consulting and Clinical Psychology* 64(4): 712–723.

Mallick, B. (2014). "Cyclone shelters and their locational suitability: An empirical analysis from coastal Bangladesh." *Disasters* 38(3): 654–671.

Marx, M., Phalkey, R. & Guha-Sapir, D. (2012). "Integrated health, social, and economic impacts of extreme events: Evidence, methods, and tools." *Global Health Action* 5: 19837. http://dx.doi.org/10.3402/gha.v5i0.19837, accessed on 8/6/2015.

Masten, A. S. (2001). "Ordinary magic: Resilience processes in development." *American Psychologist* 56(3): 227.

Mehedi, H., Nag, A. K. and Farhana, S. (2010). "Climate induced displacement- case study of Cyclone Aila in the Southwest coastal region of Bangladesh." Humanitywatch, Khulna.

Nahar, N., Blomstedt, Y., Wu, B., Kandarina, I., Trisnantoro, L. and Kinsman, J. (2014). "Increasing the provision of mental health care for vulnerable, disaster-affected people in Bangladesh." *BMC Public Health* 14(1): 1–9. www.biomedcentral.com/1471-2458/708, accessed on 17/12/2014.

Neria, Y., Nandi, A. and Galea, S. (2008). "Post-traumatic stress disorder following disasters: A systematic review." *Psychological Medicine* 38(4): 467–480.

New York Times. (2007). "Cyclone warning saved many lives." 23 November.

North, C. S. and Pfefferbaum, B. (2013). "Mental health response to community disasters: A systematic review." *JAMA* 310(5): 507–518.

Paul, B. K. (2009). "Why relatively fewer people died? The case of Bangladesh's Cyclone Sidr." *Natural Hazards* 50(2): 289–304.

Paul, B. K. (2010). "Human injuries caused by Bangladesh's Cyclone Sidr: An empirical study." *Natural Hazards* 54(2): 483–495.

Paul, S. K., Paul, B. K. and Routrary, J. K. (2012). "Post-Cyclone Sidr nutritional status of women and children in coastal Bangladesh: An empirical study." *Natural Hazards* 64(1): 19–36.

Paul, S. K. and Routray, J. K. (2011). "Household response to cyclone and induced surge in coastal Bangladesh: Coping strategies and explanatory variables." *Natural Hazards* 57(2): 477–499.

Phalkey, R., Dash, S. R., Mukhopadhyay, A., Runge-Ranzinger, S. and Marx, M. (2012). "Prepared to react? Assessing the functional capacity of the primary health care system in rural Orissa, India to respond to the devastating flood of September 2008." *Global Health Action* 5, doi:10.3402/gha.v5i0.10964

Rahman, M. (2013). "Climate Change, Disaster and Gender Vulnerability: A Study on Two Divisions of Bangladesh." *American Journal of Human Ecology* 2(2): 72–82.

Rajkumar, A. P., Premkumar, T. S. and Tharyan, P. (2008). "Coping with the Asian tsunami: Perspectives from Tamil Nadu, India on the determinants of resilience in the face of adversity." *Social Science & Medicine* 67(5): 844–853.

Sastry, N. and Gregory, J. (2013). "The effect of Hurricane Katrina on the prevalence of health impairments and disability among adults in New Orleans: Differences by age, race, and sex." *Social Science & Medicine* 80: 121–129.

Schmidlin, T. W. (2011). "Public health consequences of the 2008 Hurricane Ike windstorm in Ohio, USA." *Natural Hazards* 58(1): 235–249.

Silverman, W. K. and La Greca, A. M. (2002). "Children Experiencing Disasters: Definitions, Reactions, and Predictors of Outcomes.: In La Greca, A. M., Silverman, W. K. and Vernberg, E. M. (eds.) *Helping Children Cope With Disasters and Terrorism*. Washington, DC, American Psychological Association: 11–34.

Terranova, A. M., Boxer, P. and Morris, A. S. (2009). "Factors influencing the course of posttraumatic stress following a natural disaster: Children's reactions to Hurricane Katrina." *Journal of Applied Developmental Psychology* 30(3): 344–355.

Thienkrua, W., Cardozo, B. L., Chakkraband, M. L. S., Guadamuz, T. E., Pengjuntr, W., Tantipiwatanaskul, P., Sakornsatian, S., Ekassawin, S., Panyayong, B., Varangrat, A., Tappero, J.W., Schreiber, M. and Griensven, F. V. (2006). "Symptoms of posttraumatic stress disorder and depression among children in tsunami-affected areas in Southern Thailand." *JAMA* 296(5): 549–559.

Turpin, G. (2011). "Climate change, disasters and psychological response in Bangladesh." www.shef.ac.uk/psychology/about/news/2011/climatechange-bangladesh-turpin.

Weems, C. F., Taylor, L. K., Cannon, M. F., Marino, R. C., Romano, D. M., Scott, B. G., Perry, A. M. and Triplett, V. (2010). "Post traumatic stress, context, and the lingering effects of the Hurricane Katrina disaster among ethnic minority youth." *Journal of Abnormal Child Psychology* 38(1): 49–56.

4 Gendered health impacts of disasters

4.1 Introduction

The previous chapter focused on the physical and psychological health impacts of disasters. It explained the contexts of vulnerability during disasters and revealed the influence of the demographic factor on the health impacts of disasters. The present chapter continues this investigation through cross-sectional data analysis and explains the influences of economic, socio-cultural, behavioural and environmental factors on the health impacts of disasters.

Here, gender receives special attention, and the chapter argues that cultural attitudes, norms and traditions create and maintain differences in gender identity, roles and responsibilities and strongly influence women's vulnerability and lack of healthcare access during disasters. Poverty also makes this situation worse. The chapter concludes by explaining the complex inter-relationships of several factors and their strong impact on women's health in disasters.

4.2 Gender, a significant risk factor, and health impacts of disasters

Disasters have strong impacts on victims, but human social factors intensify the severity for each victim variably depending on their gender, socio-economic conditions (Cash et al. 2014) and location of their residence. Considering all the factors, the present research has found 'gender' to be a significant factor creating differences among victims and becoming very complicated when considered alongside culture, traditions and social attitudes. A common theme uttered by the respondents was that 'women are more vulnerable than men in all aspects of disasters'. Alam and Rahman (2014) mentioned in their research that 'during disaster a kind of functional disorder gets created where women had to face challenges different from men' (Alam and Rahman 2014, p. 68; Alam and Collins 2010; Rahman 2013), and some cultural differences between men and women contribute to the disproportionate impacts on women (MacDonald 2005), which can also be noticed in the higher number of female deaths and injuries in most disasters (Neumayer and Plümper 2007; Alam and Collins 2010; Rahman 2013). Four times more women than men were killed in the tsunami-affected areas of Indonesia, Sri Lanka and India (Oxfam 2005 cited in MacDonald 2005), which Doocy mentioned in his research: 'the resulting female deficit will likely be the tsunami's most deeply felt and prolonged impact' (Doocy et al. 2007, p. 273). The present research shows that reactions to traumatic events and disaster-related mental problems were found among both male and female victims, whereas females were noticed to suffer more (Anderson and Manuel 1994; Jacob et al. 2008; Terranova et al. 2009; Nahar et al. 2014). But unfortunately, women 'also suffer from inverse care law after disaster' (Alam and Rahman 2014, p. 68) in regards to accessibility to healthcare facilities, reliefs and social support, which become aggravated in poor socio-economic backgrounds.

Why are women more vulnerable?

If the discussion starts from the preparation period in regards of getting the warning information and evacuating the house for the shelter, it is revealing that in most cases, female members could not receive the updated information on the cyclones (CCC 2009; Rahman 2013; Nahar et al. 2014) and had to depend upon the male members for the information and decisions regardless of all socio-economic conditions (Ikeda 1995; Alam and Collins 2010; Alam and Rahman 2014). Cultural practice and conservativeness of the society do not allow women to participate in the public space (CCC 2009 cited in Alam and Rahman 2014) and create a higher dependency of women on men to get warning information (CCC 2009), especially when TV, radio and mobile networks become unavailable before the cyclone attack. Preparedness for disaster requires decision-making and leadership, and women are generally excluded from such roles in Barguna (Alam and Collins 2010; Alam and Rahman 2014).

Badar's description highlights the situation of the female members in a family for the information and decision dependency: 'We (male members of the family) went to the main roads to watch . . . I became very afraid when I saw the water with a high current entering into the locality through the roads . . . Then I went back to house and told my family to take shelter in the neighbour's two-storied house . . .' (Male, age 45, urban area). Badar's family included five children, a nursing mother and other adults. They were all depending on his information and his decision to take shelter. They had to run for their lives at the last moment.

Again, Rokeya described her situation in Tentulbaria village during Sidr. 'A strong wind started blowing . . . My sons went to see the condition of the river and they came back shouting, "all women must go to a safe place". I took my two daughter-in-laws with me and set off . . . but where should we go?' (Female, age 65, rural area).

These situations were common when the men had taken the decision to evacuate and directed their family at the last moment (Alam and Collins 2010). Being dependent and getting a sudden decision to evacuate, the female respondents were in a state of confusion, dilemmas and panic. They were not confident that they could reach a safe shelter during the disaster with their children because of their lack of knowledge of the locality. Jasim in his interview mentioned the situation for women during cyclone warning periods. 'Men can reach a safe place quicker than women . . . Women remain in their houses and do not get a chance to know the information about the locality which all the men know . . . So women get lost' (Male, age 20, urban area).

Women in rural Bangladesh mostly remain inside their houses (Alam and Collins 2010; Alam and Rahman 2014), and they have limitations on going outside the houses in regards to age, place and time, depending on the approval of the family and society. Women in Barguna (both earning and non-earning, students and housewives) 'spend more time at home' or 'at the work/study place' and have limitations on their movement which prohibit them from knowing more about the locality. During my fieldwork, I met a family where two sisters and one brother live with their parents in the centre of the Barguna municipality. I saw that the younger brother of six years roamed around the whole neighbourhood after school, but his sisters, age 11 and age 15, were always accompanied by someone, even when going for their tuition. Their parents did not feel it was safe to let their girls go alone. The reason behind this limitation is not only the fear of violence but also losing a good reputation in society. This cultural practice binds women, especially unmarried women, inside their houses and keeps them under continuous pressure (Rashid and Michaud 2000; Alam and Rahman 2014). The research findings of Naomi Hossain (2008) show 'the unquantifiable psychological costs young women bear from protecting their reputation while also trying to get ahead in school or work, with the mobility and public interaction' (Hossain 2008, p. 1).

Not only these local girls, but I myself as a female researcher did not even feel safe and comfortable to travel within the study area alone. Local people do not accept any woman moving without company. As a new woman, the problems intensified for me. I was even being teased while passing the bazaar areas. So, most of the respondents, especially young women, prefer to take company with them while going outside the home, even during the day. Sara, a young government officer in Barguna, mentioned in her interview that 'I always take Munni (young female neighbour) with me when I go outside home . . . bazaar, shopping or a visit to the doctor' (Female, age 24, urban area). And at night it becomes mandatory, as Raj mentioned in his interview: 'a woman going outside home alone for her own shopping is not positively accepted by the society. Even, a woman cannot go outside home in the evening to collect necessary medicines without male company' (Male, age 25, urban area).

These social practices 'limiting time and place for the women' created constraints for the female respondents to know about their surroundings and collect the necessary information before any cyclone. They avoid public places like tea stalls or restaurants at the bazaars or any gatherings, from where men got updated news warning when electricity was to be cut or mobile networks off. Women had to rely on men to get the latest information and the decisions for taking safe shelter in disasters, which led them to panic. During the interviews, when most male respondents mentioned their plans to take shelter during a disaster attack, many female respondents described their helplessness due to limited information about secure shelters during a sudden cyclone. Besides, the long distance to the cyclone shelter and their inability to fight against strong winds and storm water, their lack of life-saving skills like swimming (Rahman 2013) and climbing trees, coupled with their mother's protective instincts (Alam and Collins 2010) to save their children (Alam and Rahman 2014) and feeling of responsibility for their houses and household possessions make women confused. Similar reasons of vulnerability and death were revealed among the 2004 Asian Tsunami affected women in Indonesia, Sri Lanka and India (MacDonald 2005).

Again, leaving necessary goods at home unlocked (as most poor houses do not have a good locking system) might mean losing everything. Paru, a poor woman, lives in Asrayan, and she described to me that when she and her neighbours left the house during Mahasen in 2013, she lost almost everything. 'We did not have anything left . . . Young boys came around . . . They took the only fan I have, all the kitchen utensils . . . as much they could take . . . looted' (Female, age 30, urban area). In the interviews, both the male and female interviewees mentioned that women delay or sometimes refuse the call of evacuation the first time.

Faruk from the rural area described how his mother was in a dilemma whether to take shelter, 'My mother did not want to leave the house, she thought if she would die, she wanted to die in her own house . . . I had to push her to save her life . . . I carried her all along to the neighbour's house' (Male, age 48, rural area).

A female respondent, Sara, described the situation: 'My mother usually does not want to leave the home and her belongings . . . She cannot leave because of her responsibilities, though she knows her vulnerability' (Female, age 24, urban area).

Weakness, malnutrition and the fight for life

In addition to the lack of information, weaker bodies and less strength due to malnutrition and ill health (Ahsan and Khatun 2004 cited in Paul 2010; Neumayer and Plümper 2007) also make female victims vulnerable. Most female respondents mentioned in their interviews that they feel physically weak against the strong force of disasters compared to men, and they lack skills, like swimming or climbing trees. They become very weak in a short time because they are underweight or suffering from malnutrition due to inadequate intake of food throughout their lives (Haque et al. 2014). The neglect of women in regard to food distribution among family members is a well-established fact in Bangladesh, especially in rural society (Rahman 2013). They are last in taking food in amount and quality in the family hierarchy, and this is intensified in conditions of poverty (Haque et al. 2014, p. 1). Son priority, especially considering 'sons' as family security for food and health (Ray-Bennett et al. 2010), creates these gendered differences in food distribution in the family. This tradition and social attitude result in ill health and malnutrition among one-third of Bangladeshi women (Ahmed and Ahmed 2009), and anaemia among 70% of women in Bangladesh (Haque et al. 2014, p. 1), weakening them physically in the face of environmental hazards (Rahman 2013). Samsun has been a Government Family Welfare Assistant for 21 years and works with women from different socio-economic statuses in Barguna. She told me about the women's situation in regard to food consumption.

> Women do not take care of themselves . . . They do not take food in a timely fashion . . . give most food to the males . . . They first serve the food to the adult male members, then to the children . . . and only then to themselves . . .

But they give less to the girls . . . We advise them to give more food to the girls as they have deficiency . . . but they neglect the girls more . . . place high importance on boys.

(Female, age 48, urban area)

This 'son priority' is explained by Paru, a female respondent remembering her childhood. 'Sons will do any job . . . take care of parents but daughters will go to another's house . . . They will eat there . . . but is it possible for them to get enough food in the other house [Husband's house]? They [parents] do not understand that' (Female, age 30, urban area).

But these situations become more intensified for women in poor families. Maksuda: 'Women in poor families do not have much food left after serving the children . . . They do not eat . . . They get gastric . . . They have leftovers mostly . . . Men will not eat leftovers . . . Woman become more sick' (Female, age 38, urban area).

Though a few respondents mentioned that people are becoming aware of the need for equal food distribution between girls and boys, the discrimination and the prejudice lie deep in social norms. From my own experience during the fieldwork, I had the opportunity to have meals (breakfast, lunch, dinner and snacks) with local families. I attended their daily meal on invitation. From this, I saw that in both in rich and middle-class families, girls eat less, and women eat the least. They may have enough rice, but the most nourishing portions of food (protein) are always for the men and boys. Girls are even forbidden to ask for more food. I remember one invitation to a rich family's meal where I was served a big piece of chicken, breaking the normal culture. All the women were having dinner at the table, but I saw they were mostly having a very small amount of chicken, including a young girl sitting beside me. I asked her mother why she is not having the same as me, as I knew there was some left. Her mother smiled and did not answer, but the smile showed her helplessness and sadness. The answer came from the young girl: 'I'd like that big piece of chicken . . . but should I have it? . . . [smiled like mother] . . . It is always for him [pointing to her brother]'. Other senior women at the table laughed at me as if it was a silly question to be asked. I really felt sad for that girl and all other girls of Barguna where they do not get the food they need even if the family can afford it.

Pregnancy and vulnerability

Gynaecological problems and pregnancy with their weaker body added more sufferings for the female victims during cyclones. They were mostly treated as the other family members, and there are no special plans or arrangements for pregnant woman or nursing mothers to take shelter safely during the cyclones. Sometimes their problems are ignored by family members, as mentioned by Papri from Barguna municipality: 'We took my pregnant sister-in-law to the neighbour's building with us. There was no special plan for her . . . we planned, while taking shelter we would take her with us' (Female, age 38, urban area).

But this is not appropriate for pregnant women and new mothers because when other members can struggle against strong winds or water and can run fast or jump or swim, these women may get injured, lose their child or even their lives. This is unsatisfactory in view of the advice of 'having an emergency plan may reduce negative effects of disaster on the health of postpartum women and their infants' (Zilversmit et al. 2014, p. e83).

Maksuda from the urban area described the problems created for two pregnant women. Rima ended up with miserable memories, losing her child and sustaining a serious injury (Case Study 4.1).

Also, new mothers were at high risk.

Fulon: One of my sisters was living in Asrayan . . . and gave birth to a baby in the morning. Sidr started that day . . . water entered the room . . . Her father-in-law said that he would take her to his neighbour's house . . . They started to go . . . but a tree fell on her father-in-law's leg; he lost that leg . . . Then a tree fell on my sister's head . . . I lost my sister . . . but the baby survived (Female, age 45, urban area).

Saila: During Sidr my aunt . . . a new mother . . . was sick . . . She was hurt and her stitches split while jumping over fallen trees (Female, age 28, urban area).

Unfortunately, these experiences of Cyclone Sidr have not in any way improved the conditions for pregnant women and new mothers. No special plan, no arrangements and no special attention were mentioned by the respondents for pregnant women during Cyclone Mahasen, six years after Cyclone Sidr. A similar result was revealed among postpartum women in Arkansas. Only about 48% of postpartum women reported having an emergency plan for disasters (Zilversmit et al. 2014).

'Saris' and social attitude

Most women in Bangladesh, especially in remote areas wear a 'sari' (a long cloth of 6m) as their dress. In Barguna, almost all married women wear a sari, and many wear a Borka (an overall above the dress). When wearing a sari, swimming or jumping over fallen trees is very difficult. It is said to be a 'death noose' for women in strong winds or water because they easily get tangled with trees or any other objects. Most female respondents mentioned that this unfriendly dress (Alam and Collins 2010; Rahman 2013; Alam and Rahman 2014; Nahar et al. 2014) and their long hair were sources of tension for them, creating difficulties when taking shelter. They suffer, get injured and even die during cyclones (Alam and Rahman 2014). Rima, from Tentulbaria village, lost her sister during Cyclone Sidr, who was tangled with her long hair, could not release herself and was drawn into the running storm water. Rima was help-less watching her sister dying. Again, Asma (another interviewee) was injured while running in a strong wind and being stuck by her dress. Many women in

Barguna were found dead after Cyclone Sidr, hanged or tangled with their sari or with their long hair.

Beside the dress itself, maintaining a woman's dress is an issue of social honour and reputation which is expected from women even during disasters (Rashid and Michaud 2000; Rahman 2013), making them more vulnerable. Comparatively speaking, clothing is not a problem for the men. Most wear a 'lungi', a cloth about 2m long, or sometimes pants. A lungi is easy to take off when necessary. During previous cyclones, men have used this opportunity to save themselves, mentioned by most respondents. Men did not have to think of losing their honour or being humiliated. Fulon, a female respondent from the urban area, mentioned her feelings and worries in regards of maintaining her dress during the cyclones.

> A woman cannot run . . . A man can run and even take off their clothes . . . Women think, I have children, my honour and the honour of my husband . . . People may tease him (the husband) after disaster, "your wife ran on the disaster day, taking off her clothes" . . . Whereas men take off their clothes and run for their lives.
>
> (Female, age 45, urban area)

Not only Fulon, but most female respondents emphasized that women try to maintain their dress all the time, even against the elemental forces of a cyclone. They would rather die than take off their sari, though they would still have a blouse and skirt on underneath. They are especially afraid of living with humiliation. During different meetings, I found that women's dress becomes an important topic to be mentioned by all the respondents. They mentioned several incidents where women faced humiliation for losing their dress during disasters. Rokeya expressed her high concern about women's dress in disasters: 'when they were rescued, most of them [victims] did not have any clothes on . . . even many of the women . . . Some covered themselves with banana leaves . . . Some women took the urna of unmarried young girls to cover themselves' (Female, age 65, rural area). Similar findings were revealed in Dominelli's research on the Sri Lankan tsunami 2004 victims, as she mentioned, 'Death is preferred over dishonour, which of course has implications not only for the woman or girl concerned, but also for her family' (Dominelli 2013, pp. 77–93).

Here should be mentioned that the salwar-kamiz, which is mostly a young girl's dress in the rural area, is more convenient in disasters. But it is not always accepted for wearing by women of all ages.

Responsibility for children

During cyclones, both parents take responsibility to save the children and take them to a place of safety. But it becomes most challenging for the mother to save the children because of her own vulnerabilities. The situation is risky for both of them, mother and child. Many causalities were created during Sidr in saving children. Respondents mentioned several incidents of mother and child who struggled for life together but failed. Dead bodies were found after disasters where mothers were holding their children. Paru, a young mother of three girls, described her own miserable struggles to save her child.

> When we were at the main road . . . the water reached us . . . We started swimming . . . My daughter was very young . . . she could not walk . . . I took her with my one hand . . . and I started swimming with another hand . . . Oh! What a strong current . . . Can I leave my daughter? . . . Should I kill her? . . . I took her . . . I tried my best . . . At last I reached dry land.
>
> (Female, age 30, urban area)

But conditions were not favourable for all. Rima described her helplessness to save her younger brother. She could not but lose her brother that day. During the interview, it was very difficult to listen to her experiences of the cyclone. Rima's experiences are elaborately described in Case Study 4.1.

Case study 4.1

Rima is a young woman of 28 years and lives in Tentulbaria village. She could not finish her primary school education, and, at present, she is a housewife in a poor family. She is the second wife of a senior man who has grown-up sons and daughters. She did not have her own child after the Sidr incident, when she lost the first child. While visiting Tentulbaria village, I met Rima. She came to see me and talk after a long walk from her home. I saw a slim woman wearing a sari and borka (overall above the sari). Her presence made an excitement among the other villagers present at that time. They were very eager to introduce her to me because she is among one of the lucky survivors, alive to tell her true story. I was also very eager to meet with her because I have been advised by many of my respondents to take her interview. I have started the interview after introducing myself and making her feel at ease with me. She started her story, 'That day [the day of Cyclone Sidr in 2007] I did not want to go to my father's house . . . I was sick [advanced pregnancy] . . . but they requested as there was an invitation for all . . . I went to my father's house . . .'. Reaching her parents' house, Rima found that the whole family was getting prepared for the party; there was a party mood, and women were singing and having fun when they were preparing pittas. However, the warning for Cyclone Sidr was forecast, but her family avoided the warnings. In her words, 'They did not believe warnings . . . they took it as usual . . . they believe cyclone would not happen'. The party went well; invitees came on time at lunch time and enjoyed the food.

After the party, in the late afternoon, Rima received a call from a relative from another village that she should leave that area and took preparation with the young children of the family for the approaching cyclone. Finding herself in danger, Rima requested the relative to bring the trawler (large wooden boat) to take her home. But the relative was helpless, as it was very difficult and unsafe to travel with the trawler at that time. So, Rima felt trapped and became upset. She realized that she was one of the most vulnerable members of the family if the cyclone really attacked, as she was pregnant at an advanced stage, and she did not know how to swim. The whole family started to panic. The rain started, and the strong wind of Sidr started to blow, and Rima started crying. The family was not prepared at all to face the strong Cyclone Sidr. After some time, the huge trees of the house were toppling down and breaking. One of them fell on the house's kitchen and damaged it.

Rima started screaming with fear. Rima's brother was planning to cut the ropes of the eight cows, but at that time, the huge storm water entered the house. All the children, about seven in number, were lying on their beds, which suddenly floated. All the family members started crying, screaming and panicking. They all took shelter at the first floor of the wooden tin shade house, but the water reached up to that level. Rima saw death. In her words, 'I started crying, I was sure I was going to die . . . I was telling my father what would happen to me, everybody would swim but what should I do. . .'. Rima's father managed to get some empty plastic barrels for Rima and her cousin's sister to hold, as they did not know how to swim. The water current and wind were so strong and so sudden that in a second, the whole house was taken away by the cyclone. Within three or four hours, the party house had become a house of horror and death. Rima explained, 'I do not know what happened to the others . . . if they could manage to escape or not . . . how did they manage . . . but I was in the water, floating', but Rima found herself floating with her little brother, six to seven years old. They were both floating and swallowing water, as they did not know how to swim. They were trying to hold anything floating to keep them above the water. But it was very tough for Rima with her physical condition, as she was heavy with the baby and holding her brother. Rima described the situation: 'my brother was saying, "sister I am swallowing water . . ." but what should I do . . . after sometime I left my brother in the water . . . I was dying and I could not able to save my life . . . I let my brother into the water . . . I do not know where he went'. Rima lost her brother. She could not save her little brother. During the interview, it was very tough to hold the tears to see the sister expressing her helplessness to save her brother. I saw an inexpressible sadness in her eyes. I saw her feeling guilty for saving herself rather than saving her brother. What a difficult decision Rima had to take on the night of Cyclone Sidr.

After some time, Rima found herself alone in the dark, floating in the water. She could not see or find her mother, father, brother or anyone from the family. All she could see were the trees and floating objects. She was getting hurt and cut by those objects. She was calling for help, but there was no one. After some time, while passing a tree which had leaves with thorny ends, Rima realized that it was a Palmyra tree. The leaves were hurting her. Rima managed to climb on to the tree to save herself from drowning.

She tried to remember any prayer (Sura or Dua) but was so distressed that she could not remember any. She was calling for help. While waiting on the Palmyra tree, Rima had to face another loss of her family member. Her sister with long hair was stuck with the lower part of the tree trunk, but she could not release her hair and died after struggling with the strong water. Rima could not do anything but see her sister dying in front of her. Two sisters were struggling to live, but the same tree saved one and killed the other.

After some time sitting on the tree, she felt labour pains. She felt helpless. Rima described, 'The pain started . . . the baby . . . I was alone, my sister was dead under the tree . . . I was alone. . .'. As she was hurt by the trees and different objects while floating in the water, the labour pains started earlier than the due time. After some time, she became senseless and the birth process was half done, and the baby was not born alive. Rima remained senseless for the rest of the night on the top of the Palmyra tree. Next morning, as she remembers, she became conscious when her brother-in-law was crossing under the tree, crying loudly in grief. Her brother also noticed that there was a line of blood flowing from the top of the Palmyra tree, and as the surge water had reduced, he was standing on land; it became too high to see the top of the tree. He called other people and climbed the tree. He was surprised to see Rima there and was more astonished to find her alive. They planned to get Rima from the tree, but there was no easy way, no hope for help from outside of the affected village, and it was impossible to think of any emergency service in that remote area of Bangladesh after a devastating cyclone. So, they planned to tie pipes around her body to make a stretcher structure and pushed her from the tree into the small pond under the tree. Being hit by the water, the baby came out from the mother, as he had been stuck halfway for the whole night. Rima fainted again. The umbilical cord could not be cut properly and had to be torn by hand by the female neighbour. She was carried to the nearest school and kept lying on the tables, wrapped with papers. Rima lost parts of her clothes, and there was no cloth available to cover her. Rima lost all her family members that night except her father. Rima's father found her the next morning at that school when he had lost hope of finding anyone alive from his family.

The school was full with victims, and they were all crying and grieving for their lost nearest and dearest ones. Mothers were crying for their lost children, sisters were crying for brothers and sisters and injured victims were crying in pain. Children were crying for hunger, thirst and losing their parents.

Rima was injured all over her body. She was exhausted and very tired. She was very thirsty. But there was no fresh water for her, and the villagers had to give her the dirty water from the puddles. She waited at this stage until her husband came in the afternoon and took her to the nearest healthcare centre at Phuljury. They gave her stitches in the wound and medicine and kept her at the centre. But the bleeding was not stopped with the local treatment, and she needed further medicine from outside. She received the medicine from a group of visitors who came to visit the area after Sidr. Rima needed to stay at the hospital for one month. But unfortunately, she could not have a baby again, as she needed a complete treatment, which is costly for her. She went to Barguna and got medicine from the non-qualified medical providers of Barguna municipality, but she needed medical treatment. She could not visit Barisal for the proper treatment due to lack of money, though there is another issue behind this delay. Rima is the second wife of a senior man, and he already had grown-up children, so he is not going to spend more money for Rima to get the full treatment, as he is not eager to have any more children. But Rima wants her own child, and Sidr has ruined her dream, and economic dependency is creating a delay. She is a poor and helpless victim of Cyclone Sidr, but no help (from the family or out-side) reached Rima to complete her treatment to recover the severe damage created in her life.

This case shows the situations created during Cyclone Sidr for the pregnant women in the rural areas of coastal Bangladesh and their vul-nerability to the health impacts of cyclones due to the lack of gender-sensitive disaster management plans considering the social attitude, behaviour and norms. Rima represents most of the poor women living in disaster-prone regions. They need special help and attention; their sto-ries get published in the newspapers but do not get priority in research or in disaster response and recovery plans.

Post-disaster additional problems for women

Post-disaster periods bring additional health problems for women. Injured female victims face several barriers to receiving healthcare which lower their accessibility to injury treatments compared to male victims. These healthcare conditions and determinants will be elaborated in the next chapter of this book. Besides injury treatments, damaged sanitation is a problem for the women after disasters, which becomes intensified during their pregnancy or menstruation due to the socially negative attitudes towards them (Rashid and Michaud 2000; CCC 2009). In the present research, most of the women from all socio-economic conditions faced difficulties where the sanitation system was damaged at home (CCC 2009; Rahman 2013). This condition was described by Azhar, an NGO officer who works for victims of disasters. 'After disasters when the cyclone becomes weaker . . . women, especially young women become vulnerable due to lack of sanitation, the lack of places for taking baths, the unavailability of clean clothes and the cultural attitudes in regards of all these' (Male, age 38, urban area).

Lack of food and relief collection

'Women suffer from every side of disasters – health, cleanliness, food and cultural norms'. This is a quote from Azhar, an NGO official, while explaining the conditions of female victims after disasters.

> Gender discrimination could be seen higher after disaster . . . A young boy can run and take relief from the streets but a young girl or her mother may be starving but cannot go for food. They have to wait until the father or brother brings the food.
>
> (Male, age 38, urban area)

Rokeya described the situation. 'When a man brings relief food we all eat . . . and if there is an old woman like me . . . we go because children cry being starved' (Female, age 65, rural area).

But, 'every woman cannot go to the street to collect relief . . . I could not', Maksuda mentioned in her interview (Female, age 38, urban area). Women's socio-economic conditions and family reputations also hinder women queuing for relief. Female victims from the middle class or better-off families avoid collecting relief because of the socially negative attitudes. Runa, a female respondent, explained in her interview, 'My husband went to collect the relief . . . Women do not beg, they feel ashamed to request relief. I do not have the habit to ask for anything' (Female, age 30, rural area).

Increased household works

Gender differences in workloads brought additional suffering to the female respondents after disasters (Fothergill 1998; CCC2009; Rahman 2013; Alam and Rahman 2014). Cooking daily food for a whole family in a damaged kitchen with a shortage of utensils, water and food items is a great challenge. Besides, cleaning mud or clay from the floor, bringing water from distant sources, taking care of the children and helping male members to repair the damage, all of this work is done by the female members, which is sometimes a great burden.

Papri, a female respondent from a middle-class family in Barguna municipality, described the suffering she faced after Sidr.

> There was no electricity for one month . . . or a water supply. I suffered bringing water from long distance . . . At night I had to carry 20–30 buckets of water, fill the store and work the whole day with that water. During normal times I do not have to do that. I even had to take clothes to wash and kids for bathing to the distant location . . . I became sick.
>
> (Female, age 38, urban area)

Respondents also mentioned that due to these increased workloads for the women during the post-disaster periods, female victims could not visit healthcare centres.

Men, vulnerability and privileges

Men's vulnerability, like other members of society, was also created while taking shelter during the disaster with no preparation and sometimes taking risks for the duties that are socially expected of them (Doyal 2000). In the present research as well other research, men were reported to have been injured while saving their family members and neighbours during Cyclone Sidr (Alam and Collins 2010). No male left their family in a vulnerable house without calling them first, as all respondents reported. Kamrul, a male respondent, said that 'no man can leave their female family members in a disaster, whatever happens, he will take them. If there is any child in the family, how could they leave them?' (Male, age 28, rural area). But sometimes, men had to leave their families behind to save themselves because they could run fast, and saving others became a great challenge. Jasim explained the situation: 'men first tried to save female members but when they could not . . . they left them to save themselves. Everyone loves their own life' (Male, age 20, urban area).

Besides the responsibility of the family, valuable assets like cattle, goods of a shop and houses become a great burden for the men during cyclones. Many men did not leave their vulnerable houses to protect the cattle or valuable assets until the situation became severe (Alam and Collins 2010; Paul 2010).

After reaching the shelter or confirming their family's safety, many men and sometimes women tried to rescue other victims of their locality. But mostly, saving neighbours and bringing them to a safe shelter was the voluntary effort of local men. They risked their lives to save floating persons in the water. They helped neighbours take their cattle to safe shelters and rescued disabled persons and pregnant women from the storm water. These activities sometimes brought injuries or even cost them their lives. The following are quotations to illustrate these points.

Maksuda: My brother got hurt rescuing an old paralyzed woman from her affected house (Female, age 38, urban area).

Liza: All the men got involved in rescuing the victims whereas no woman did this (Female, age 35, rural area).

Psychological health impacts and gender

Gender is revealed as an important factor for vulnerability to the mental health impacts of disasters. A traumatic event itself is the most powerful determinant of the psychological harm, and no person is absolutely resistant to the severe attack of disasters (Herman 1997). Beyond personal traits, women's vulnerability in disasters is mostly influenced by the dominant social and cultural attitudes (Norris et al. 2002). A quotation of a female victim highlights the mental condition of women in regard to facing cyclones: 'When I hear warnings, I feel very scared . . . I feel insecure . . . Men can fight back, they know they can do it . . . but a woman has to consider lots of things' (Female, age 30, rural area). As mentioned in the previous sections, women felt difficulties due to their weaker bodies, lack of life saving skills and knowledge and especially their dependency on men for information and decisions. Again, pregnancy and responsibilities for children also create difficulties (Nahar et al. 2014). All of these situations make them scared and depressed. Most of the time, no special arrangements were made by the family; they were treated like any other family member. Maksuda described in her interview that 'there were two pregnant women in the house . . . They were crying as others were getting prepared to run in the strong wind . . . but they could not do that . . . and they were afraid for their lives' (Female, age 38, urban area). Again, most female respondents faced intense tension about their children and panicked when the cyclone warning came, as described by Papri.

> I took two of them. My son was with his aunt . . . We started running to the nearest building . . . no sandals . . . no preparation . . . we reached the building . . . people and people . . . all in that building . . . but my tension was whether my son reached . . . was he outside the building?
>
> (Female, age 38, urban area)

Again, a disabled, sick old woman was left helpless with her old husband in a vulnerable house during the disaster by the other family members. Later, she was rescued by a kind neighbour, but the experience was not good for her: 'she was crying a lot as she was left behind by her own family . . . She became a burden for the neighbours . . . She was rescued from certain death' – described by Maksuda (Female, age 38, urban area), who was present at that time. Here, Herman's quotation highlights the situation created for women in disasters: 'traumatic events, like other misfortunes, are especially merciless to those who are already troubled' (Herman 1997, p. 60).

Male victims also mentioned their fear, anxiety and helpless conditions facing cyclones. They mostly fear for the safety of their children and other family members. They also became very worried to save their valuables. But compared to female victims their vulnerability is not intensified by socio-economic and cultural factors.

There is an inspiring mental condition revealed among many Cyclone Sidr victims irrespective of gender. Both male and female victims of Cyclone Sidr tried to face the disaster bravely and save themselves and others. Though men were mostly engaged in rescues, many women also showed their bravery by saving babies and helping others. Even pregnancy or a physically weaker body did not prevent them from these courageous steps. Again, the confidence to face future cyclones was revealed among both men and women. Though a higher number of men were found to be more confident and planning to face a future cyclone as decision-makers in the family, women also mentioned their self-belief in the interviews. Thus, for Runa, 'My life is also very important . . . and I am ready to fight back to cyclone attack . . . if necessary I would do anything to survive' (Female, age 30, rural area). As a member of a patriarchal society with a subordinate position, Runa's confidence is really inspiring.

Gender differences, discrimination and violence

During the interviews, respondents were asked if they faced any gender discrimination in taking shelter or staying safe from the impacts of disaster. Most respondents mentioned that women were not forbidden to take shelter in the cyclone centres among men, and they did not face direct gender differences in evacuation, as mentioned in some research (see Alam and Collins 2010). But gender difference is imbedded in the social attitudes and culture well before any disaster attacks. Barguna and NGO officials also mentioned an increase in gender discrimination and violence after disasters, which will be highlighted in Chapter 7.

Men and women start the life-saving race at the same time during cyclones, but men mostly reach safety first. Sara (Female, age 24, urban area) mentioned that '[m]en can run fast but women cannot', and 'men have cultural privileges in facing disasters'; this was the common opinion mentioned by most of the respondents in their interviews. While physical strength helped men in testing conditions, the cultural attitude of 'freedom to go anywhere' was the key for them to know the location of safe shelters and the optimum routes to get there. They had the freedom to take their own decisions; for instance, they could shorten their dress, whereas this was impossible for women to contemplate. As Maksuda explained, 'men ran for life with less clothes, even naked in disasters but women took risks to keep their saries on them . . . or they would be humiliated afterwards' (Female, age 38, urban area). Most females and even male respondents mentioned that this cultural attitude towards woman had created a higher number of casualties in Cyclone Sidr, which was also a factor in the higher women's casualties in the 2004 Asian tsunami (MacDonald 2005).

With regard to violence, no occurrence was mentioned by cyclone evacuees in Barguna, whereas violence was common in the overcrowded camps of tsunami victims in 2004 (MacDonald 2005; Pittaway et al. 2007). However, women in the Barguna shelters were often at high risk of verbal abuse and sometime physical harassment (Neumayer and Plümper 2007). Evacuees in the cyclone shelters had to be very alert, and all the local residents looked after each other, especially the female members. Mintu explained, 'When we (some the local residents) went outside the school (cyclone shelter), we requested the other men to look after our families and they did . . . and when they needed, we also did the same for them . . . We, all men were alert that time . . . We ourselves took different measures' (Male, age 52, urban area). Similar risks and occurrences of verbal abuse have been revealed among flood victims in Bangladesh (Rashid and Michaud 2000) as for the tsunami victims in South Asian countries (MacDonald 2005).

4.3 Socio-economic condition and health impacts of disasters

Complex socio-economic factors (Marx et al. 2012), such as gender (Nasreen 2004), educational attainment of the household head (Karim et al. 2014) and a household's access to human, financial, natural, physical and social assets have a strong influence on the intensity of cyclone victims' problems (Webster 2013; Karim et al. 2014) and sacrifices they have to make, such as reducing investment in child nutrition, healthcare and education after disasters (Skoufias 2003). But all over the world, it is primarily the poor who have suffered during the recent natural disasters (CCC 2009; Rahman 2013), and injuries were also found to be higher among the poor (Paul 2010). Research has revealed that poor people in developing countries like Bangladesh and even in the USA become easy victims (CCC 2009; Rahman 2013). During Hurricane Katrina, the poor were mostly affected because of their residence in low-lying areas, their lack of private transport and even the lack of proper rescue attention to them (Jacob et al. 2008).

Vulnerability during disaster attacks

The present research reveals that economic conditions make the poor people's Barguna residences more vulnerable to facing challenges during disaster attacks. The stronger houses of the better off did not need to be evacuated during the cyclone, though their location did work negatively in some cases, like in Tentulbaria village. In general, strong houses helped respondents to save their cattle and valuables, and some even had stores of dried foods for the worst days of the post-disaster period (Ray-Bennett et al. 2010). These strong houses also created opportunities for neighbours, relatives and friends to take shelter. Liza, a school teacher and a member of a rich family, explained, 'We did not leave the house; it is a brick house and is strong. People came from neighbouring areas to take shelter at our house . . . we welcomed them' (Female, age 35, rural area).

Houses of the poor in Barguna are very weak against the strong winds of the cyclone, and their locations are also very vulnerable to storm surges. Mainly, this ultra-poor group lives on the river bank, being landless. Even the Government project for landless people, 'Asrayan', is just beside the river. These inhabitants have to suffer even during normal high tides, as river water enters their houses. So, they know about their own vulnerability and have a strong awareness after Cyclone Sidr. Sometimes, they wait or take shelter in the nearest strong house. But poor respondents, especially the new migrants/refugees, do not have good connections with rich neighbours and have smaller social networks. If new to the area, they may not even have close relatives in the locality. Similar conditions have been reported by Hurricane Katrina refugees (Wilson and Stein 2006 cited in Bonanno et al. 2010). Smaller social networks seem to correlate with longer delays in evacuation during disasters. There is a scarcity of cyclone shelters, and they hesitate to take shelter until emergency conditions are upon them. These poor inhabitants become helpless. Their situation is illustrated by Fulon's interview.

Researcher:	Where did you stay during Mahasen?
	Fulon(Female, age 45, urban area): I was at my house. . .
Researcher:	Were you scared to be in a Gol pata house? [A house made of leaves, 'Gol Patta', a very vulnerable poor house]
Fulon:	Water entered into my house.
Researcher:	Did not you take shelter anywhere then?
Fulon:	I went to my brother's house . . . It is near . . . My son, who is 7 years old, said, 'ma we should not stay at this house . . . let's go. . . .' I started running with him . . . A betel-nut tree fell . . . a papaya tree fell . . . Then we started screaming and ran fast to that house . . .

Cyclone Mahasen was more frightening for Fulon, though it did not have much impact in Barguna compared to Cyclone Sidr. The reason was explained by her.

'Why should I be scared in Sidr? We four families took shelter in our relative's strong house after we heard the warning by the hand miking . . . When the daylight was still there . . . we were all together in that room . . . Water did not enter that much . . .'. Fulon was at her village that time. Sidr damaged her whole house, and she could not repair it because of a lack of money and had to leave the house. At present, she is living in Barguna municipality and works as a maid in another's house to maintain her family. This condition shows that being in her own village, she at least had relatives on hand and was in a safe condition during Sidr. She could at least take shelter during Sidr, but during Mahasen, she took risks for herself and her young child.

Post-disaster suffering

Post-cyclone resilience was also influenced by socio-economic status. Better-off and well-networked victims received more help from relatives in emergency funds and emotional support. In their turn, they are then able to help the less fortunate (Alam and Collins 2010; Ray-Bennett et al. 2010) with food, clothes and necessaries to face the devastation. Maksuda, for instance, was able to have meals the day after Sidr, as her parents and sisters came forward to help her. 'After the disaster we had nothing left to eat . . . but the next day dry foods were provided from my father's house and my sister sent me cooked rice' (Female, age 38, urban area).

Respondents from better-off families were able to cook their own meals in the badly affected Tentulbaria village the day after the cyclone. Their stronger houses enabled more cooking options. They also had stored fuel, such as oil or dry wood, and stored rice and lentils. After cleaning up and repairing any damage, they were able to return to some semblance of normality and have some cooked food at home, whereas 'poorer households are typically less equipped to deal with shocks' (Skoufias 2003, p. 1088), and most of the poor respondents lost everything and became dependant on outside relief or help from others. Interviews with the respondents revealed that especially the ultra-poor victims had to rely on cooked relief food, dry processed food with limited nutrition (Islam et al. 2011, Karim et al. 2014) and other necessary items like clothes, groceries and medicine. But unfortunately, relief was not timely (Islam et al. 2011) or equally distributed among the victims. Insufficient relief, delays in distribution, along with corruption and mismanagement created a difficult situation for the poor to survive.

Runa, mother of a young child, explained her feelings. 'The next morning my son started asking for food . . . he wanted rice . . . where I would get rice? . . . There was nothing left . . . Fallen trees were everywhere . . . Water covered whole area . . . Where could we go for rice? . . . I did not have any dry rice to cook' (Female, age 30, rural area).

Question: Did no relief aid reach you that day?
Runa: Yes, they gave us puffed rice, sweets alongside the bank of river . . . Later my husband went to get them.

Unfortunately, supplied cooked food from the local authority did not reach all of those affected in the study area.

Relief aid and help: the only means to survive

Relief aid was one of the main supports for the victims after Cyclone Sidr, especially the adversely affected poor inhabitants. Except in some areas, most of the relief aid reached the affected areas two to three days late. But when it did arrive it helped the victims a lot, as described by Rokeya, one of my senior respondents (Female, age 65, rural area) from the rural areas who remembered the cyclone in 1970 and compared it with Cyclone Sidr. 'We suffered a lot in the last cyclone [1970] but in Sidr we got lots of things . . . clothes, food'.

A summary of the interviews shows that different things like cooked foods (Kichuri), rice, dried rice, puffed rice, banana, bread, sweets, lentils, edible oil, flour, sugar, biscuits, drinking water, medicine (water purification tablets, worm tablets, oral saline), clothes and sanitary pads were distributed as relief items. But among the respondents, different opinions were expressed in regard to the amount and type of relief items and also the mismanagement and corruption in relief distribution. They reported that most of the relief items were insufficient in amount per person or per family. Sifa from Tentubaria village mentioned, 'There was no medicine in the relief items, only water purification tablets and oral saline' (Female, age 50, rural area). Salam said, 'All of them did not get relief . . . only 5% of the victims received water purification tablets' (Male, age 19, rural area).

Though relief was planned to be distributed among the poor, there were complaints that the poor or affected people were not included on the list of recipients.

Corruption and mismanagement among staff, distributors, local powerful people and the NGOs in relief distribution created helplessness among the victims. Sometimes, real victims were hindered or prohibited to receive help and support. Respondents were keen to talk about relief distribution.

Mintu described in detail with anger and an upset voice the conditions created among the poorest people of the Barguna municipality in Asrayan.

> People who live in Asrayan are very poor . . . if any relief comes, it should reach Asrayan first but people of Asrayan do not receive anything . . . It was all gone before reaching Asrayan (during Cyclone Sidr). All the relief was grabbed at the administration level . . . All the relatives of local authority officials got relief . . . They declared that they were penniless whereas they were in fact very rich, they had lots of money, they had 20–25 cows . . . Actually our relief was controlled by them.
>
> (Male, age 52, urban area)

Sahida (Female, age 35, urban area) said, 'The well-off got more but those without got nothing . . . The Chairman and Members repaired the houses of those who would give them bribes'.

Beside these corruptions and mismanagement, sometimes, the way the relief effort was allocated did not help the poorer victims. The targeting of the relief programme was based on the size of asset losses experienced by households rather the level of household assets prior to the cyclone, which helped the better-off victims receive support. Similar experiences were observed in Honduras after Hurricane Mitch in 1998. There was only limited scope for providing more relief to those who suffered greater losses or who were poorer following a disaster (Morris and Wooden 2003 cited in Skoufias 2003, p. 1097).

Repairing damage

Again, repairing the damage from Cyclone Sidr became a huge burden for the victims, especially the poor respondents (CCC 2009), because they did not have much capacity. They barely received any help from formal credit services, and other sources of financing were often inaccessible, which prevented them from replenishing their stocks of productive assets (Skoufias 2003). They had to suffer for long periods, with some still living in damaged houses six years later.

I met an old woman in Asrayan who is still living in her damaged house. She does not have enough money for repairs, and her suffering was described by her daughter.

> My mother's house was damaged in Sidr . . . the roof leaks . . . Rainwater enters . . . She has to wake up at night when it rains and sit to one side and again when the rain stops she goes to sleep . . . What can she do? She does not have any money . . . My father died and did not leave enough property.
>
> (Female, age 30, urban area)

Psychological impact and socio-economic statuses

Socio-economic status is always associated with the levels of mental distress during disasters (Norris et al. 2002), and 'lower socio-economic condition was consistently associated with greater post-disaster distress' (Norris et al. 2002; Nahar et al. 2014). Present research reveals that victims of poor and middle-class economic conditions are more vulnerable to mental impacts of disasters, which is also revealed among the Hurricane Katrina survivors who are still living in the temporary houses (Harrison 2007 cited in Jacob et al. 2008). Again, the poor are systematically under-served by the country's mental health services in Bangladesh (Nahar et al. 2014). It should be mentioned that good socio-economic conditions do not always protect rich victims from the psychological impacts of cyclones. A safe, strong house, economic solvency to provide emergency needs after disaster and social/family support to arrange basic needs and injury treatment during and after disasters did not prevent some from panicking during Cyclone Mahasen. They mentioned several mental impacts such as panic attacks, anxiety, intense fear and even the symptoms of post-traumatic stress disorder. For them, the memories of Sidr and the loss of family members made them scared during Mahasen's warning period.

At the other end of the response spectrum, there is evidence of resilience and the confidence to fight against future cyclones among respondents irrespective of socio-economic condition. Poor victims are affected more in disasters, but for many, their confidence and resilience against future cyclones are excellent and inspiring. For instance, Tentulbaria village faced extreme devastation from cyclone Sidr due to its vulnerable location and being an underdeveloped area, but the mental strength of the inhabitants was not much different from the respondents of Barguna municipality and surrounding areas. They are afraid of cyclones but at the same time are confident of their abilities. Their powers of coping are advanced.

4.4 Poverty, gender and impacts of disasters

Socio-economic conditions and gender have a strong relation to the impacts of disasters on health. Previous sections have revealed the problems of poor victims during disasters, but considering gender shows an intensification of problems for poor women. Poor women in Barguna were found to be facing more physical and psychological problems during cyclones that resemble those in other research. Neumayer and Plumper, for instance, revealed after analyzing disaster effects in 141 countries over the period 1981 to 2002 that 'when the socio-economic status of women is low, more women than men die' (Neumayer and Plümper 2007, p. 552), and Macdonald in her research on tsunami victims mentioned that 'when natural disasters hit poverty-stricken areas, women are more likely to be affected than men' (MacDonald 2005, p. 178). During the Kobe earthquake in 1995, 1.5 times as many women as men died, and many elderly women died because they lived in poor residential areas, which were more heavily damaged (Neumayer and Plümper 2007, p. 555). Poor women become the most disadvantaged among the disadvantaged groups.

Poor women's problems start from the warning period, when they are encouraged to leave their homes. They face various problems during evacuation, as elaborately described in the previous sections. But reaching a cyclone shelter is not the end of their problems; rather, they are intensified. 'Cyclone shelters are not female friendly' – this was the common opinion among the interviewees in the present research (CCC 2009; Alam and Rahman 2014). Overcrowding is common in all the cyclone shelters, with no separate arrangements for newborn babies, new mothers and pregnant women (CCC 2009; Haider and Ahmed 2014; Nahar et al. 2014). There is no separate sanitation facility for the women (CCC 2009). Staying in a cyclone shelter for some hours means a loss of dignity, which motivates women from different socio-economic classes or families to avoid such shelters (CCC 2009). The present research revealed that female members of the middle class or solvent families avoid overcrowded cyclone shelters and are wary even of sheltering in a rich neighbour's house. Not all from the middle class have strong houses, but none of them evacuated in Sidr. The reasons are social (Mallick 2014) and quite complicated. Position in socio-economic groups, family values and heritage, overcrowded environments in the cyclone shelters and conservativeness (not religious conservativeness) influence women not to leave their houses. They try to face their problems within their own capacity, which makes them potentially vulnerable during disasters.

Again, when collecting food and other items from the relief lines was essential for the poor respondents, gender was a strong filter. Interview analysis shows that women (mostly young women and girls) were counted second among the family members who went to collect the relief. There were some families where female members suffered at home without the proper amount of food and other items after Cyclone Sidr rather than going to the relief lines (Alam and Rahman 2014). Women from all families and all ages are not expected in the queues, whereas men do not have these restrictions. Culture, tradition and the gendered attitude of the society are behind this. Lack of food and unequal gendered distribution in the family and society intensified the food scarcity among the female victims (CCC 2009; Paul et al. 2012).

These sufferings of women become more exaggerated in female-headed families or with the loss or absence of male members. As women need help in every aspect after disaster, poor female-headed households suffered, along with single mothers (husband left or did not take care of them) and widows (Rahman 2013; Alam and Rahman 2014; Morrow and Phillips 1999; Zilversmit et al. 2014). During interviews, several female respondents mentioned that they needed help from a man for economic support, relief collections, repairing the house and security, and they became very helpless if they did not have support from their father, brother or husband. Paru, a single mother of three girls, lives in Asrayan, explained her helplessness during Cyclone Sidr:

> I did not have my husband with me, so I am helpless . . . he has a mental disorder and I do not know where he is . . . I cannot maintain my family with three daughters . . . After Sidr what sufferings I had to face . . . I did not have a house or cash, but I have three daughters.
>
> (Female, age 30, urban area)

4.5 Conclusion

In conclusion, this research illustrates in detail that, to quote Bonanno et al. (2010, p. 1), 'there is no one single dominant predictor of disaster outcomes'. Several inter-related social, cultural and economic factors create complicated situations for each victim, and the result is impacts of different intensities. The research shows how this follows lines of existing disadvantage and inequality. In this regard, this chapter has elaborately discussed the context and reasons for disaster health impacts on victims, followed by cross-sectional data analysis of demographic, behavioural, economic, socio-cultural, health and environmental factors (some demographic and behavioural factors are described in Chapter 3). The analysis reveals that all of these factors are inter-related and work in complex ways, influencing causations and health impacts of disasters (Figure 4.1).

Among the factors, economic conditions – income, employment and possession of resources – affect the availability of strong houses for shelter and money for emergency treatment after disasters. Better-off people do not need to evacuate their houses, which saves them from the health hazards of evacuation, and they are stronger in the face of the post-disaster emergency periods, whereas poor inhabitants are always vulnerable due to their weak houses, need for shelter and lack of money in disaster periods. Social factors, including family and social support, are very important in this regard, especially during evacuation, rescuing victims and surviving the difficult period post-disaster. This social support is the only means to survive for many poor victims until any relief aid reaches the affected area.

Other environmental factors, like local cyclone risks, severity of the cyclone's effects and location of the residences, affect the whole process, safe evacuation and availability of better transport systems, shelters and relief. Relief aid reaches urban areas earlier than the rural areas, where the economic conditions of the inhabitants are comparatively poorer. So, the vulnerability of people to the health impacts of disasters depends on their actions taken in disasters, resulting from this interplay of different inter-related factors.

Again, the chapter highlights that when we consider gender and its relation with other socio-cultural factors, it is clear that the health impacts of disasters are worse for women than men. Analysis shows that male and female victims are influenced by common environmental, demographic, behavioural and economic factors, but cultural attitudes and norms create differences in gender relations and responsibilities that seriously affect the health conditions in disasters, increasing the vulnerability of women (Figure 4.2).

Dependency for disaster information and evacuation decisions, responsibility for children and household things, lack of life-saving skills and unfriendly dress and factors affecting health, including malnutrition and pregnancy, create difficult situations for the female victims while taking shelter during cyclone attack. Their health conditions are also affected during the post-disaster periods due to increased household work and responsibilities, the stigma of relief aid collections and using sanitation. All these impacts of gendered attitudes of general society become more pronounced for poor women due to their weak houses and dependency on cyclone shelters and the collection of relief. Therefore, the

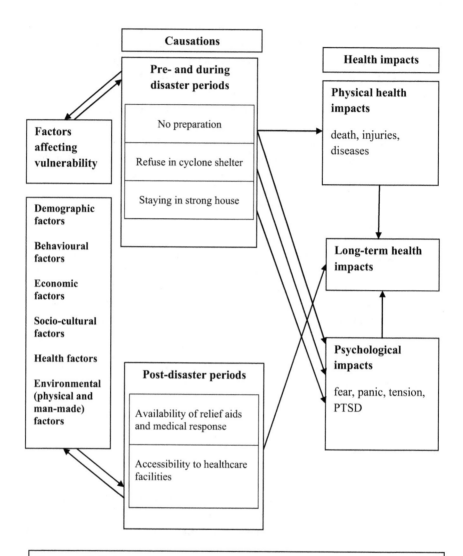

Demographic factor: age
Behavioural factors: unawareness, ignorance to weather warnings, access to healthcare
Economic factors: income, employment, possessions of resources
Socio-cultural factors: gender, family and social support, cultural attitude, norms and traditions
Health factor: health status; sickness, malnutrition and pregnancy
Environmental (physical and manmade) factors: severity of cyclone, Local cyclone risk,
Location: urban and rural, availability of facilities; safe shelters, medical responses, healthcare facilities

Figure 4.1 Key relationships among factors, causations and health impacts of disasters

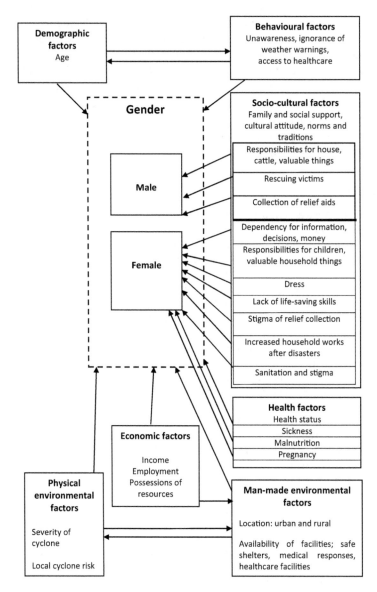

Figure 4.2 Relationships between gender and vulnerability factors of disaster health impacts

inter-relation of poor economic condition and socio-cultural factors results in making the poor women the most vulnerable members of the society compared to men. Thus, this research, by taking an intensive qualitative approach, has drawn out detailed aspects of culture, and in so doing it demonstrates why these larger patterns exist. The complex inter-relations of these demographic, socio-cultural, economic, health and environmental factors have been mentioned in other research (Enarson 2002; Skoufias 2003; Keim 2006; Few and Tran 2010; Karim et al. 2014; Webster 2013; Bolin et al. 1998; Turner et al. 2003 and Anwar et al. 2011). But without the close analysis of victims' experiences that has been discussed in this chapter, it is not always clear exactly where the problems lie nor how they might be resolved.

The next two chapters move on to investigate the healthcare access of the victims, especially women, after disasters.

References

Ahmed, T. and Ahmed, A. (2009). "Reducing the burden of malnutrition in Bangladesh." *BMJ* 339: b4490.

Ahsan, R. M. and Khatun, H. (2004). *Disaster and the Silent Gender: Contemporary Studies in Geography*. Bangladesh, Geographical Society.

Alam, E. and Collins, A. E. (2010). "Cyclone disaster vulnerability and response experiences in coastal Bangladesh." *Disasters* 34(4): 931–954.

Alam, K. and Rahman, M. H. (2014). "Women in natural disasters: A case study from southern coastal region of Bangladesh." *International Journal of Disaster Risk Reduction* 8: 68–82.

Anderson, K. M. and Manuel, G. (1994). "Gender differences in reported stress response to the Loma Prieta earthquake." *Sex Roles* 30(9–10): 725–733.

Anwar, J., Mpofu, E., Mathews, L., Shadoul, A. F. and Brack, K. E. (2011). "Reproductive health and access to healthcare facilities: risk factors for depression and anxiety in women with an earthquake experience." *BMC public health* 11(1): 523.

Bolin, R., Jackson, M. and Crist, A. (1998). "Gender Inequality, Vulnerability, and Disaster: Issues in Theory and Research." In Enarson, E. P and Morrow, B. H. (eds.) *The Gendered Terrain of Disaster: Through Women's Eyes*, New York, Praeger: 27–44.

Bonanno, G. A., Brewin, C. R., Kaniasty, K. and Greca, A. M. L. (2010). "Weighing the costs of disaster consequences, risks, and resilience in individuals, families, and communities." *Psychological Science in the Public Interest* 11(1): 1–49.

Cash, R. A., Halder, S. R., Husain, M., Islam, M. S., Mallick, F. H., May, M. A., Rahman, M. and Rahman, M. A. (2014). "Reducing the health effect of natural hazards in Bangladesh." *The Lancet* 382(9910): 2094–2103.

CCC. (2009). *Climate Change, Gender and Vulnerable Groups in Bangladesh*. Dhaka, Climate Chamge Cell, DoE, MoEF, Component 4b, CDMP, MoFDM: 1–82.

Dominelli, L. (2013). "Gendering Climate Change: Implications for Debates, Policies and Practices." In Alston, M. and Whittenbury, K. (eds.) *Research, Action and Policy: Addressing the Gendered Impacts of Climate Change*. London, Springer: 77–93.

Doocy, S., Rofi, A., Moodie, C., Spring, E., Bradley, S., Burnham, G. and Robinson, C. (2007). "Tsunami mortality in Aceh province, Indonesia." *Bulletin of the World Health Organization* 85(4): 273–278.

Doyal, L. (2000). "Gender equity in health: Debates and dilemmas." *Social Science & Medicine* 51(6): 931–939.

Enarson, E. (2002). "Gender Issues in Natural Disasters: Talking Points on Research Needs." *Crisis, Women and Other Gender Concerns*. ILO, Geneva, Working paper, 5–12.

Few, R. and Tran, P. G. (2010). "Climatic hazards, health risk and response in Vietnam: Case studies on social dimensions of vulnerability." *Global Environmental Change* 20(3): 529–538.

Fothergill, A. (1998). "The Neglect of Gender in Disaster Work: An Overview of the Literature." In Enarson, E. P. and Morrow, B. H. (eds.) *The Gendered Terrain of Disaster: Through Women's Eyes*, New York, Praeger: 11–25.

Haider, M. Z. and Ahmed, M. F. (2014). "Multipurpose uses of cyclone shelters: Quest for shelter sustainability and community development." *International Journal of Disaster Risk Reduction* 9: 1–11.

Haque, M. M., Bhuiyan, M. R., Naser, M. A., Arafat, Y., Roy, S. K. and Khan, M. Z. H. (2014). "Nutritional status of women dwelling in urban slum area." *Journal of Nutritional Health & Food Engineering* 1(3): 1–4.

Harrison, E. (2007). "Suffering a slow recovery." *Scientific American* 297(3): 22–25.

Herman, J. (1997). *Trauma and Recovery: The Aftermath of Violence – From Domestic Abuse to Political Terror*. New York, Basic Books.

Hossain, N. (2008). "The price we pay." *FORUM, a Monthly Publication of The Daily Star*, 3(1). www.thedailystar.net/forum/2008/january/price.htm, accessed on 22/9/2014.

Ikeda, K. (1995). "Gender differences in human loss and vulnerability in natural disasters: A case study from Bangladesh." *Indian Journal of Gender Studies* 2(2): 171–193.

Islam, A. S., Bala, S. K., Hussain, M. A., Hussain, M. A. and Rahman, M. M. (2011). "Performance of coastal structures during Cyclone Sidr." *Natural Hazards Review* 12(3): 111–116.

Jacob, B., Mawson, A. R., Payton, M. and Guignard, J. C. (2008). "Disaster mythology and fact: Hurricane Katrina and social attachment." *Public Health Reports* 123(5): 555–566.

Karim, M., Castine, S., Brooks, A., Beare, D., Beveridge, M. and Phillips, M. (2014). "Asset or liability? Aquaculture in a natural disaster prone area." *Ocean & Coastal Management* 96: 188–197.

Keim, M. E. (2006). "Cyclones, tsunamis and human health." *Oceanography* 19(2): 40–49.

MacDonald, R. (2005). "How women were affected by the tsunami: A perspective from Oxfam." *PLoS Medicine* 2(6): e178.

Mallick, B. (2014). "Cyclone shelters and their locational suitability: An empirical analysis from coastal Bangladesh." *Disasters* 38(3): 654–671.

Marx, M. et al. (2012). "Integrated health, social, and economic impacts of extreme events: Evidence, methods, and tools." *Global Health Action* 5: 19837. http://dx.doi.org/10.3402/gha.v5i0.19837, accessed on 8/6/2015.

Morris, S. S. and Wodon, Q. (2003). "The allocation of natural disaster relief funds: Hurricane Mitch in Honduras." *World Development* 31(7): 1279–1289.

Morrow, B. H. and Phillips, B. (1999). "What's gender 'got to do with it'?" *International Journal of Mass Emergencies and Disasters* 17(1): 5.

Nahar, N., Blomstedt, Y., Wu, B., Kandarina, I., Trisnantoro, L. and Kinsman, J. (2014). "Increasing the provision of mental health care for vulnerable, disaster-affected people in Bangladesh." *BMC Public Health* 14(1): 1–9. www. biomedcentral.com/1471–2458/708, accessed on 17/12/2014.

Nasreen, M. (2004). "Disaster research: Exploring sociological approach to disaster in Bangladesh." *Bangladesh e-journal of Sociology* 1(2): 1–8.

Neumayer, E. and Plümper, T. (2007). "The gendered nature of natural disasters: The impact of catastrophic events on the gender gap in life expectancy, 1981–2002." *Annals of the Association of American Geographers* 97(3): 551–566.

Norris, F. H., Friedman, M. J., Watson, P. J., Byrne, C. M., Diaz, E. and Kaniasty, K. (2002). "60,000 disaster victims speak: Part I: An empirical review of the empirical literature, 1981–2001." *Psychiatry: Interpersonal and Biological Processes* 65(3): 207–239.

Oxfam. (2005). "The tsunami's impact on women." www.oxfam.org.uk/what_we_do/issues/conflict_disasters/downloads/bn_tsunami_women.pdf, accessed on 3/5/2005.

Paul, B. K. (2010). "Human injuries caused by Bangladesh's Cyclone Sidr: An empirical study." *Natural Hazards* 54(2): 483–495.

Paul, S. K., Paul, B. K. and Routrary, J. K. (2012). "Post-Cyclone Sidr nutritional status of women and children in coastal Bangladesh: An empirical study." *Natural Hazards* 64(1): 19–36.

Pittaway, E., Bartolomei, L. and Rees, S. (2007). "Gendered dimensions of the 2004 tsunami and a potential social work response in post-disaster situations." *International Social Work* 50(3): 307–319.

Rahman, M. (2013). "Climate change, disaster and gender vulnerability: A study on two divisions of Bangladesh." *American Journal of Human Ecology* 2(2): 72–82.

Rashid, S. F. and Michaud, S. (2000). "Female adolescents and their sexuality: Notions of honour, shame, purity and pollution during the floods." *Disasters* 24(1): 54–70.

Ray-Bennett, N. S., Collins, A., Bhuiya, A., Edgeworth, R., Nahar, P. and Alamgir, F. (2010). "Exploring the meaning of health security for disaster resilience through people's perspectives in Bangladesh." *Health & Place* 16(3): 581–589.

Skoufias, E. (2003). "Economic crises and natural disasters: Coping strategies and policy implications." *World Development* 31(7): 1087–1102.

Terranova, A. M., Boxer, P. and Morris, A. S. (2009). "Factors influencing the course of posttraumatic stress following a natural disaster: Children's reactions to Hurricane Katrina." *Journal of Applied Developmental Psychology* 30(3): 344–355.

Turner, B. L., Kasperson, R., Matson, P. A., McCarthy, J. J., Corell, L. C., Christensen, L., Eckley, N., Kasperson, J. X., Luers, A., Martello, M. L., Polsky, C., Pulsipher, A. and Schiller, A. (2003). "A framework for vulnerability analysis in sustainability science." *Proceedings of the National Academy of Sciences* 100(14): 8074–8079.

Webster, P. J. (2013). "Meteorology: Improve weather forecasts for the developing world." *Nature* 493(7430): 17–19.

Wilson, R. K. and Stein, R. M. (2006). *Katrina Evacuees in Houston: One-Year Out.* Unpublished manuscript, Houston, TX, Rice University.

Zilversmit, L., Sappenfield, O., Zotti, M. and McGehee, M. A. (2014). "Preparedness planning for emergencies among postpartum women in Arkansas during 2009." *Women's Health Issues* 24(1): e83–e88.

5 Impacts of disaster on healthcare accessibility

5.1 Introduction

Injury is a major cause of death and a primary cause of morbidity in tropical cyclones (Meredith and Bradley 2002; Keim 2006). After disasters, rescuing patients and providing medical attention in a timely manner are very important (Johnson and Galea 2009) to save victims' lives and mitigate aftereffects. Within response and transportation times, victims at risk of death in the first few hours of disasters will likely perish before reaching healthcare centres (Johnson and Galea 2009; Fawcett and Oliveria 2000), and delays, inadequate treatment or mismanagement of injuries may lead to infections and ultimately to disabilities (Phalkey et al. 2011). Structural elements such as emergency transportation and patient transfer and burdens on the health system in arranging the provision of services become a great challenge after disasters (Johnson and Galea 2009), especially in developing countries, where weak healthcare systems are already overloaded (Phalkey et al. 2011). Beside these structural elements, socio-demographic, socio-psychological, dynamics of illness and social factors such as gender and socio-economic conditions influence the victims' healthcare-seeking behaviours after disasters (Uddin and Mazur 2014). Inhabitants of Barguna have faced several factors in accessing healthcare during cyclones. Among them, insufficient local healthcare facilities and disrupted transport systems, lack of emergency services and external medical help were common, whereas socio-economic conditions and gender-created inequalities among the victims made the poor and women, especially poor women, more vulnerable in receiving timely healthcare after disasters. All these factors influenced victims' health and healthcare access from the disaster night to long after the cyclone. The present chapter focuses on environmental factors, such as availability of healthcare facilities, emergency services, relief and accessible transport, and behavioural factors, whereas the next chapter highlights gender as a socio-cultural factor and analyzes the influences of other factors: demographic, economic and health factors. Before going into detailed descriptions and analyses of these factors, we will look at the consequences of these factors on injured victims as a context for the following discussion.

Interview analysis revealed that no injured respondent received healthcare immediately (in the first 12 hours) after the incident. Even critical, life-threatening injuries received medical attention only after 20 hours. A few collected medicine from a pharmacy the next day without consulting a medical doctor. The first and main healthcare providers for injuries were the local pharmacy and drugstore salesmen. Later, injured people sought medical attention from local healthcare centres (government hospital, sub-centres and private consultation rooms) after five to six days when their health problems became intolerable (Biswas et al. 2006), but some victims remained untreated. Similar findings were revealed in other research by Parvin et al. (2008) and Uddin and Mazur (2014), where rural, poor cyclone victims did not seek treatment until their illnesses became severe and paramedics, pharmacy and drugstore salespeople were found to be the major healthcare providers for Cyclone Sidr's victims in the coastal zone of Bangladesh.

The impacts of delayed and improper treatments were found to be negative among the respondents in the present research. Minor injuries were reported to be healed with prolonged pain and discomfort, but most major injuries became health disabilities for the victims (Phalkey et al. 2012; Rathore et al. 2012). Kamal's experience of losses during Cyclone Sidr highlights this point. 'My brother got hurt on his leg . . . he was taken to the medical centre 7/8 days after Sidr after difficulties in walking . . . for that period he took only paracetamols from the pharmacy . . . but due to lack of proper treatment timely, his leg became weak . . . he has got a disability now . . . he is only 37 years old' (Male, age 48, rural area). Again, many injuries did not receive complete treatment because of poor socio-economic conditions and lack of external financial help.

Detailed descriptions and analyses of these factors will be helpful to highlight the impacts of disasters on healthcare access in the remote districts of Bangladesh, where insufficiency in health facilities is a major pre-existing problem.

5.2 Insufficient local healthcare facilities and accessibility

Bangladesh has an inequitable distribution of healthcare resources, especially in remote areas. More than 60% of people in Bangladesh have no access to modern healthcare services other than immunization and family planning (Rahman 2005 cited in Ray-Bennett et al. 2010). Research reveals that problems in the government sector lowered its healthcare services to only 14% of all instances of illness (Ahmed et al. 2005), forcing patients to use private practices and the informal sector. Remote districts like Barguna face insufficiency in both public and private sectors, not a promising context in view of the correlation in Bangladesh between the concentration of qualified health workers and key health outcomes (Ahmed et al. 2011). Vacancies in medical professional posts, high rates of absenteeism among medical doctors and lack of diagnostic facilities and medicine are all common problems of the low-cost public healthcare sectors in Barguna, as well as other areas of the country (Health Bulletin 2013; Ahmed et al. 2005), and the private sector in Barguna is also so undeveloped that it is unable to provide emergency treatment. The inhabitants have to rely on the only Government Hospital located in Barguna municipality, or they have to visit the healthcare centres of the divisional cities outside Barguna (Hasan 2012). Every emergency cases need long-distance travel, except for those within the municipality.

After a disaster such as a cyclone, because of the disrupted transport system, it becomes a big challenge for the patients to travel from their remote locality and visit the only Government Hospital located in the municipality or one of the distant sub-centres. This sparsity of conveniently located facilities restricts injured victims in receiving any health care after cyclone disasters. Remote rural areas like Tentulbaria village in Barguna (Ahmed et al. 2011; Hasan 2012) suffer most. There is no healthcare centre or pharmacy in Tentulbaria, and in normal times, patients visit Barguna municipality or the nearest sub-centre at Phuljury, about 5km from the village. There is no land transport system, only local boats, which are not suited to carrying critical patients. Villagers can only collect common medicines from the drugstores located in nearby villages. But during disasters, most of the local healthcare centres and pharmacies in the municipality and rural areas remain closed or damaged, as happened in Cyclone Sidr, except for some areas of Barguna municipality. The situation was particularly difficult in Tentulbaria, as all of the nearest drugstores were damaged by Sidr, and, due to disrupted transport, injured victims had to remain untreated for long time, even without antiseptics and medicines.

What happens on the disaster night, just after injuries occur?

'It was impossible to get any treatment on the disaster night' was the most common experience mentioned by my respondents. A few injuries were treated with indigenous methods or antiseptic cream if it was available in the house.

Jasim (Male, age 20, urban area) described his situation. 'The branch of a tree fell on my uncle's head . . . We put chewed betel leaf on his wound . . . He was kept at home like this on the disaster night . . . The next day we collected medicine from the pharmacy'.

Using indigenous treatment methods is common in remote areas of Bangladesh (Parvin et al. 2008; Rahmatullah et al. 2010) and at times may be the only available option for injuries. Betel leaves used on skin cuts was one; others include Banana leaves, 'Balia' leaves, grass paste, tooth powder, 'Andaw' leaves, and, for broken arm or legs, home-made bandages and splints.

Nazem (Male, age 28, urban are) commented in his interview that 'rural inhabitants . . . when they do not get any treatment, they collect the leaves from the woods, paste and use them for a wound or for broken legs or arms, they use sticks and clothes to make a bandage'.

During the interviews, Barguna respondents mentioned that during previous cyclones, they had to rely on indigenous methods or antiseptic creams, especially for minor injuries, as they did not have any option, though most of them felt the need for proper medical treatment.

Again, during the night of Cyclone Sidr, thousands of people in Barguna took shelter at the cyclone shelters, and they stayed there for the whole night. Some were seriously injured while taking shelter, but there was no healthcare facility. Not even first aid, antiseptic cream or even indigenous methods were available at the urban and rural cyclone shelters. Sick people had to be in the overcrowded environment without any special arrangements. Respondents described the situation.

'We were 800/900 people in the cyclone shelter . . . no emergency healthcare facility was present in the shelter . . . We are lucky we are alive' (Male, age 48, rural area).

So, all injuries, from life-threatening to minor cuts (whether the injured party stayed at home or in the cyclone shelter), remained untreated until the morning after Cyclone Sidr.

Access or no access to treatment of injuries

Residential location made a huge difference in healthcare accessibility the day after Sidr. At the national level, about 75% of qualified physicians practise in urban areas (Hasan 2012). As in other areas of Bangladesh, Barguna municipality has more healthcare centres compared to the surrounding rural areas. The residents of Barguna municipality and surrounding areas could collect some medicines or receive treatments, but they had to walk a long distance and somehow navigate roads damaged or blocked with fallen trees. Meanwhile, rural victims remained without any health facility the day after Sidr.

Samsun (Female, age 48, urban area): 'Healthcare facilities were about 50% available in the urban area but there was nothing in the rural area . . . 0% . . . no doctor . . . the pharmacies were swept away by the storm surge . . . all medicines . . . even the windows and doors of the pharmacies were gone'.

Nazem (Male, age 28, urban area): 'The victims who live near the city could have collected any medicine but the rural victims who live far from the city had to wait until the roads became accessible . . . It was two or three days until the fallen trees were removed . . . For these days they had to suffer at home'.

According to the respondents, rural victims became trapped in their village without any treatment for two to three days before any transport become accessible or any healthcare could reach their villages. Rokeya from Tentulbaria village explained the condition of the injured victims in the rural areas. 'They [injured victims] could not do anything . . . Their injuries remained untreated for 2–3 days until relief arrived . . . There was no doctor . . . no pharmacy at the village bazaar . . . Even the bazaar itself had totally vanished . . . all swept away by the cyclone' (Female, age 65, rural area).

Among all of the rural injured respondents in the present research, only one respondent was taken to the healthcare centre, and others did not seek treatments for their major injuries, which were serious in some cases. The urban victims had more options, as they had the Government Hospital, more healthcare centres and pharmacies, and they could reach the centres by walking. But even they were restricted, one of the major reasons being the limited capacity of the local healthcare centres against the huge demand after disasters.

Limited capacity of the local healthcare centres

Most of the critically injured victims could not receive complete treatment from Barguna and were referred to metropolitan or divisional cities such as Dhaka or Barisal due to the lack of different types of injury treatments and specialist physicians (Hasan 2012). Barguna normally suffers from a severe shortage of healthcare providers, as the ratio is one government physician for 28,810 inhabitants (Health Bulletin 2013), like other remote areas in Bangladesh (Mahmood et al. 2010; Ahmed et al. 2011). The health worker (physician, nurse and CSBA) density is only 0.13/1,000 population in Barguna, which is similar (0.14/1,000 population) to Chakaria, a remote rural area of Bangladesh (Mahmood et al. 2010). These ratios fall far short of the WHO-recommended health worker ratio of 2.5/1,000 population, confirming that Bangladesh has a critical shortage of healthcare providers (WHO 2006). But after disasters, operational local healthcare centres face exceptional demand (Fawcett and Oliveira 2000; Johnson and Galea 2009) because of the damage to others (Paul et al. 2012). During Cyclone Sidr, Barguna Government Hospital and a few sub-centres, such as the Phuljury Union sub-centre, became the only hope of receiving any treatment. But increased dependency on the few open centres coupled with a severe shortage of health workers and equipment created a 'bottleneck' (Berggren and Curiel 2006) to providing proper healthcare services to the injured survivors (Paul et al. 2012). Barguna Government Hospital usually has a severe shortage of beds and health workers, which became a much greater challenge after Cyclone Sidr. The hospital has only 50 beds and nine doctors working against 21 sanctioned posts, with no female doctors and 23 nurses against 37 sanctioned posts (Health Bulletin 2013). The hospital quickly became overcrowded after Sidr with injured patients as well as their regularly admitted patients. Paru from the Asrayan (Government residences for landless people) took an injured victim to the hospital and faced difficulties there: 'there was no space at the hospital . . . They stopped the admission of new patients . . . They said [to] us, "you should go to other centres . . ." and they became impatient to handle a huge number of patients but where could we go?' (Female, age 30, urban area).

Paru's experience was similar to that mentioned by a New Orleans nurse during the post-Katrina period: 'the patient rooms are crowded, the staff is stressed, and there are serious supply shortages. Our standards of quality are tough to meet when the system is so strained' (Berggren and Curiel 2006, p. 1550). Metropolitan New Orleans faced an acute crisis and limited options, as only 15 of 22 area hospitals were open with 2,000 of the usual 4,400 beds, decreasing the ratio of 3.03 hospital beds per 1,000 population to 1.99 per 1,000 population (Berggren and Curiel 2006). Findings from other research also showed similar situations during different disasters. Djalali (2011) mentioned that Iran 'as a developing country in Asia' (Djalali et al. 2011, p. 30) faced a noticeable shortage of beds, blankets, triage tags, medicine and intravenous fluids during the first days after the Bam earthquake in 2003. Other researchers also discussed the shortage of resources in the healthcare system in the affected countries before and during the tsunami of 2004, in spite of an increased budget due to external aid (Kohl et al. 2005; de Ville de Goyet 2007 cited in Djalali et al. 2011). Total hospital capacity and medical resources became the major problem for providing the proper healthcare to the survivors (Berggren and Curiel 2006; Lai et al. 2003; Jacob et al. 2008).

Delays in proper treatments intensified the health problems of the injured victims, and, in some cases, the injuries became worse (Djalali et al. 2011). Most injured survivors could not travel outside Barguna because of the inaccessible transport and high cost of treatments. During the interviews, respondents mentioned their inability to take patients to the healthcare centres and regretted it because of the impact.

Not only the injured victims, but also patients with chronic conditions are separated from their healthcare providers or medications as a result of cyclones (Stratton and Tyler 2006 cited in Johnson and Galea 2009). After Hurricane Katrina, nearly 45% of hemodialysis patients missed one or more treatments in New Orleans (Hyre et al. 2007 cited in Johnson and Galea 2009). The present research also reveals that in normal times, specialist physicians travel to Barguna once in week, and patients wait for their next visit (Hasan 2012), but due to cyclones, these services are hindered. The condition of Saila, a young mother was affected by this situation after Cyclone Mahasen in 2013 (presented in Chapter 6, Case Study 6.1), represents the problems of other patients in Barguna after disasters.

5.3 Post-disaster periods: inaccessible transport, no emergency services and unreachable medical help

Like Saila's doctor, problems with public transportation became a major challenge (Johnson and Galea 2009) for cyclone victims from Barguna to travel for injury treatment during the first 24 hours after cyclone disasters. Though it has been reported that the armed forces responded immediately, providing search and rescue after Cyclone Sidr (Cash et al. 2014), in reality there was no external emergency service in either the urban or rural areas of the study areas to rescue and transfer injured patients to healthcare centres. Victims in the present research had to depend on local transport available for patient transfer.

In normal times, patients are mostly carried by man-drawn vans, rickshaws, local boats (trawlers) and 'Mahendro' (a small, three-wheeled, motorized vehicle). Few private ambulances were available in Barguna municipality, but anyway, these costly transports are rarely used, especially in life-threatening conditions. Again, ambulances cannot reach all areas because the remote coastal areas of Bangladesh do not have road networks, only waterways (Sarwar 2005). In Tentulbaria village, for instance, the only available transport system in and out is by boat on the local river, but people do need more than one form of transportation to reach their destination. Vehicles, rarely available in rural areas (Story et al. 2012), have to be hired from the local bazaars or from a river landing stage, which involves a walk of some distance on the local earthen roads. Besides, these emergency vehicles are not comfortable for the patient due to a lack of proper protection under rain and sun. So, there has to be a plan of how and when the patients will be carried to the distant healthcare centres considering the weather, the availability of vehicles and the opening hours of the healthcare centres. Tentulbaria's villagers mentioned in their interviews that transferring patients using a trawler in the Bishkali River is always quite risky, hazardous and costly. Insufficient transport is a major problem for their locality.

All of these transport systems become less accessible after cyclones. Vehicles are unavailable (Lai et al. 2003), and roads are blocked with fallen trees

(Islam et al. 2011). It becomes very difficult to walk along the local rural roads and earthen walkways, which are easily damaged or destroyed in heavy rains or cyclones (Alam and Collins 2010). The waterway also becomes risky for the local trawlers due to running floodwater. In most cases, injured survivors have to wait for days until the roads are cleared and floodwaters subside in the river. After Cyclone Sidr, most of the injured victims from the Tentulbaria could not receive any treatment for 72 hours, except for one critically injured woman, who was taken to the Phuljury healthcare centre by trawler 20 hours after the incident. Mostly victims living in the remote villages or far from the urban areas 'suffered at home', as Nazem described in his interview: 'The villages which are located far from the municipality . . . could not receive any healthcare the next day . . . All of the roads were closed . . . Trees were fallen on the roads . . . It took days to remove the trees . . . for these days, injured people in rural area suffered at home' (Male, age 28, urban area).

Due to inaccessible transport, patients could not be carried to healthcare centres after the disaster (Johnson and Galea 2009). Alauddin mentioned in his interview that he had to keep his injured mother at home after Sidr because of inaccessible transport. 'I could not bring her to the doctors . . . it was impossible to travel along the damaged road to the main bazaar [45 minutes walking distance] . . .' (Male, age 27, urban area). The situation was intensified for old people, children and pregnant women.

Asma from Asrayan raised an important point in her interview, asking 'How can an injured person visit the distant healthcare centre? They are sick . . . Medical staff should come to them' (Female, age 22, urban area). However, medical teams could not reach all the affected areas. Relief operations, which included visits by the medical teams and distribution of relief items, became a great challenge. The geographical location of dispersed river islands, damaged roads and lack of vehicles created a difficult situation for the relief agencies to go the 'last mile' and distribute the relief to the affected victims (Balcik et al. 2010). During the interviews, Government and NGO relief workers shared their experiences of how they faced the constraints to reach the affected areas after Cyclone Sidr. Relief agencies and medical teams mostly did not have any vehicle of their own to carry the bulk amount of relief and had to rent locally, which were either unavailable or costly because of the sudden demand (Balcik et al. 2010). Relief efforts were challenged by the damaged roads and shortage of vehicles (Cash et al. 2014). Most of the distant areas did not receive any help the day after Cyclone Sidr except for Barguna municipality and a few surrounding areas. Nazem from the Barguna municipality mentioned this in his interview: ' . . . medical teams and doctors were sent to the nearest affected areas the next day . . . from the Government, NGOs such as the Red Crescent and some private doctors went on their own initiative' (Male, age 28, urban area). Maksuda, who lives near the Barguna municipality, mentioned that normally, it takes half an hour to reach their area from the municipality, but now, it took two days after the cyclone. Similar situations have been revealed in other research. Islam et al. in 2011 and Alam and Collins 2010 revealed in their research on cyclones that proximity to the urban areas and conditions of transportation influenced relief distribution after Cyclone Sidr, and, despite the best efforts of well-meaning emergency programmes, it took four to five days to get relief to difficult geographical locations of cyclone-devastated areas (Islam et al. in 2011; Alam and Collins 2010). Present research has also revealed that most of the relief aid reached the affected areas from two days to 15 days after Sidr. But here it should be mentioned that relief items included some medicine, but relief teams did not have any medical staff with them. No respondent mentioned that they could have received any treatment from the medical staff of the relief teams. Badar, the key respondent of this research, informed me that 'I have never seen any medical staff visiting our area after disasters . . . The people who were injured went to the pharmacy as per their abilities' (Male, age 52, urban area). The absence of medical staff in the relief teams deprived victims of any chance to be checked and receive treatment on-site.

5.4 Medical relief items and dissatisfactions among users

When local healthcare centres were inaccessible and medical teams did not reach the affected areas, relief aid was the only hope for injured victims to receive medicines. However, dissatisfaction about medical relief items was revealed among most respondents of Barguna. Including the late arrival and absence of medical staff in the relief team, insufficient medicines and corruption of relief distributions deprived the victims of medical help.

Faruk from Tentubaria village expressed his anger and disappointment: 'No-one from our locality will say that we received any medicine (except oral saline) or any doctor visited our area . . . One team came to the Majer Chor [distant island] and they showed off distributing only oral saline' (Male, age 48, rural area). Besides, the survivors had to walk a long distance to reach the dispersed and distant locations of relief distribution camps and get help, which was very difficult for the injured victims. Jasim, a young man, expressed needs of the injured victims: 'If medical staff could kindly walk to the injured persons, then they would be greatly helped' (Male, age 20, urban area). Information on relief distribution was also improperly disseminated at the affected areas. Akmal from Tentulbaria village informed me that due to lack of information, his family did not receive any relief after Cyclone Sidr: 'We did not get medical relief because we did not get any information . . . The people who knew, went to collect the relief' (Male, age 19, rural area).

But collecting relief did not meet the needs of the victims, as they found that the types and the amounts of the medicine per person were insufficient and inappropriate. Self-targeting is recommended for successful relief distribution (Takasaki 2011), but, without considering the needs of the victims, only oral saline, worm tablets and painkillers were distributed as medicine (Paul et al. 2012). Also, as Jasim explained, 'They did not provide proper amount of facilities to the injured person . . . they gave very little . . . the amount of medicine provided was not sufficient according to the needs of a patient' (Male, age 20, urban area). These needs are heterogeneous across households (Morris and Wodon 2003), and a lack of evaluation only leads to chaos (Jacob et al. 2008). Again, relief was irregularly distributed without any time plan. Injured/sick victims had to wait for the next relief team to receive more medicine, without knowing the distribution time. This delay exacerbated their health problems. Kamal from Tentulbaria village mentioned this: 'They did not distribute enough medicine for a patient . . . a patient did not get well with that medicine . . . They had to wait without full treatment . . . had to wait for the next relief trawler to get medicine again, while their health condition became worse' (Male, age 48, rural area). Even medicines for children were insufficient. Samsun reported this: 'Children at the village suffered from diseases like diarrhoea, stomach upset . . . They had rural indigenous treatment or they had to go to distant healthcare centres . . . Medicines were not sufficiently supplied' (Female, age 48, urban area).

All of these problems related to relief distribution were intensified by corruption in relief distribution. Elite capture of relief aids by the community leaders (Takasaki 2011), mismanagement and conspiracy by the distributors exaggerated the scarcity of relief items and narrowed the opportunity of injured respondents to receive some medicines for their health problems after Sidr.

Mintu: The NGOs only helped their members, and not all of their members . . . only those who had connections. I was a member of . . . NGO. I did not get any help from them whereas other members received reliefs (Male, age 52, urban area).

In their interviews, respondents expressed their several opinions as 'anger' – 'the poorest residence we should get relief first but we did not get any' (Mintu from Barguna municipality); 'disappointment' – 'Those close to the chairman received many reliefs' (Fulon from Barguna municipality); or 'helplessness' – 'We could not do anything, we are helpless' (Rokeya from Tentulbaria village).

The consequences of the delayed treatment were not good for the victims. They mentioned their pain and suffering due to delays, and some of them suffered from critical medical condition and even injury turned into disability. Kamal's younger brother, Sahida's husband and Saila are the victims of delayed treatment. Though there is no research or survey in Bangladesh on Cyclone Sidr victims focusing on the delayed treatment and its impact on the injuries, but the findings of other research show that the lack of immediate and adequate emergency medical care onsite and within 24 hours increases the mortality rate and the long-term medical complications among the victims, for instance, in the Bam earthquake in Iran (Fawcett and Oliveira 2000; Djalali et al. 2011).

5.5 Psychological health impacts and healthcare access

Previous chapters of this book demonstrated that there were several psychological health impacts of cyclones among the victims, but mostly, they went unrecognized and untreated. Six years after Cyclone Sidr, many people have recovered from the impacts, but still some expressed their feelings about cyclones in a way that suggested symptoms of post-traumatic stress disorder. None of these psychological health problems got any proper assessment, attention or treatment in the post-disaster period, nor even in the following six years. None of the respondents went to a medical centre for their psychological health problems. These problems remained within the family, in the house. The main reason was that there was no recognition of these problems as needing or being susceptible to treatment. Generally, in Barguna, psychological problems get attention only when uncontrollable or a threat to public safety. One of the respondents explained the situation: 'People with mental problems . . . especially when they are uncontrollable and a threat to others are taken to Dhaka or Pabna directly. Before this, sometimes we take them to the local kabiraj or religious people' (Male, age 25, urban area). This quotation represents the common opinion on mental health in Barguna and explains the level of awareness about the psychological health impacts of cyclones, which easily create confusion between general fear of disasters and post-traumatic stress disorder. So, being unaware, no one received any mental health support after the cyclones, and even if they wanted to have any support, there was no appropriate medical facility in Barguna (information collected from the Civil Surgeon Office, 2014), and the private sector is also unable to provide any treatment. Actually, facilities for mental illness are poorly developed in Bangladesh as a whole (Nahar et al. 2014; Government of Bangladesh 2007). The whole country has only one 500-bed mental health hospital located in Pabna, in the north of the country, and the human resources working for mental health per 100,000 population are only 0.49 (Government of Bangladesh 2007). All of these constraints have created negative impacts on healthcare access for psychological health problems after disasters, whereas medical interventions and psychological support are highly recommended by several researchers (Markenson and Reynolds 2006; Tunstall et al. 2006; Paxson et al. 2012; Nahar et al. 2014).

5.6 Conclusion

Chapter 5 has looked at the issue of healthcare access after disasters, one of the major questions of this research. It focused on environmental factors (e.g. availability of local healthcare facilities and medical relief aid and accessibility of transport systems) and how these influence the healthcare access of victims, again drawing on respondents' personal experiences. According to the investigation, environmental and behavioural factors are common contributors to healthcare inaccessibility in disasters and have a direct impact on the health of disaster victims. Considering this important conclusion, the next chapter continues its investigations on other factors, especially where socio-cultural factors like gender and poverty create highly significant differences among social groups.

References

Ahmed, S. M., Hossain, M. A., Chowdhury, A. M.R. and Bhuiya, A. U. (2011). "The health workforce crisis in Bangladesh: Shortage, inappropriate skill-mix and inequitable distribution." *Human Resource for Health* 9(3): 1–7. www.human-resources-health.com/content/9/1/3.

Ahmed, S. M., Tomson, G., Petzold, M. and Kabir, Z. N. (2005). "Socioeconomic status overrides age and gender in determining health-seeking behaviour in rural Bangladesh." *Bulletin of the World Health Organization* 83(2): 109–117.

Alam, E. and Collins, A. E. (2010). "Cyclone disaster vulnerability and response experiences in coastal Bangladesh." *Disasters* 34(4): 931–954.

Balcik, B., Beamon, B. M., Krejci, C. C., Muramatsu, K. M. and Ramirez, M. (2010). "Coordination in humanitarian relief chains: Practices, challenges and opportunities." *International Journal of Production Economics* 126(1): 22–34.

Berggren, R. E. and Curiel, T. J. (2006). "After the storm – health care infrastructure in post-Katrina New Orleans." *New England Journal of Medicine* 354(15): 1549–1552.

Biswas, P., Kabir, Z. N., Nilson, J. and Zaman, S. (2006). "Dynamics of health care seeking behaviour of elderly people in rural Bangladesh." *International Journal of Ageing and Later Life* 1(1): 69–89.

Cash, R. A., Halder, S. R., Husain, M., Islam, M. S., Mallick, F. H., May, M. A., Rahman, M. and Rahman, M. A. (2014). "Reducing the health effect of natural hazards in Bangladesh." *The Lancet* 382(9910): 2094–2103.

de Ville de Goyet, C. (2007). "Health lessons learned from the recent earthquakes and Tsunami in Asia." *Prehospital and Disaster Medicine* 22(1): 15–21.

Djalali, A., Khankeh, H., Ohlen, G., Castren, M. and Kurland, L. (2011). "Facilitators and obstacles in pre-hospital medical response to earthquakes: A qualitative study." *Scandinavian Journal of Trauma, Resuscitation and Emergency Medicine* 19(1): 30.

Fawcett, W. and Oliveira, C. S. (2000). "Casualty treatment after earthquake disasters: Development of a regional simulation model." *Disasters* 24(3): 271–287.

Government of Bangladesh. (2007). *WHO-AIMS Report on Mental Health System in Bangladesh*. Dhaka, Ministry of Health and Family Welfare.

Hasan, J. (2012). "Effective telemedicine project in Bangladesh: Special focus on diabetes health care delivery in a tertiary care in Bangladesh." *Telematics and Informatics* 29(2): 211–218.

Health Bulletin 2013. (2013). Health Bulletin 2013, District: Barguna, Barguna Civil Surgeon Office, Barguna, Ministry of Health and Family Welfare, Bangladesh.

Hyre, A. D., Cohen, A. J., Kutner, N., Alper, A. B. and Munter, P. (2007). "Prevalence and predictors of posttraumatic stress disorder among hemodialysis patients following Hurricane Katrina." *American Journal of Kidney Diseases* 50(4): 585–593.

Islam, A. S., Bala, S. K., Hussain, M. A., Hussain, M. A. and Rahman, M. M. (2011). "Performance of coastal structures during Cyclone Sidr." *Natural Hazards Review* 12(3): 111–116.

Jacob, B., Mawson, A. R., Payton, M. and Guignard, J. C. (2008). "Disaster mythology and fact: Hurricane Katrina and social attachment." *Public Health Reports* 123(5): 555–566.

Johnson, J. and Galea, S. (2009). "Disasters and Population Health." In K.E. Cherry (ed.) *Lifespan Perspectives on Natural Disasters*, London, Springer: 281–326.

Keim, M. E. (2006). "Cyclones, tsunamis and human health." *Oceanography* 19(2): 40–49.

Kohl, P. A., O'Rourke, A. P., Schmidman, D. L. and Dopkin, W. A. (2005). "The Sumatra-Andaman Earthquake and Tsunami of 2004: The hazards, events, and damage." *Prehospital and Disaster Medicine* 20(6): 356–363.

Lai, T. I., Shih, F. Y., Chiang, W. C., Shen, S. T. and Chen, W. J. (2003). "Strategies of disaster response in the health care system for tropical cyclones: Experience following Typhoon Nari in Taipei City." *Academic Emergency Medicine* 10(10): 1109–1112.

Mahmood, S. S., Iqbal, M., Hanifi, S. M. A., Wahedi, T. and Bhuiya, A. (2010). "Are 'village doctors' in Bangladesh a curse or a blessing?" *BMC International Health and Human Rights* 1: 18. www.biomedcentral.com/1472-698X/10/18.

Markenson, D. and Reynolds, S. (2006). "The pediatrician and disaster preparedness." *Pediatrics* 117(2): e340-e362.

Meredith, J. and Bradley, S. (2002). "Hurricanes." In Hogan, D. and Burnstein, J. (eds.) *Disaster Medicine*. Philadelphia, PA: Lippincott, William and Wilkins: 179–186.

Morris, S. S. and Wodon, Q. (2003). "The allocation of natural disaster relief funds: Hurricane Mitch in Honduras." *World Development* 31(7): 1279–1289.

Nahar, N., Blomstedt, Y., Wu, B., Kandarina, I., Trisnantoro, L. and Kinsman, J. (2014). "Increasing the provision of mental health care for vulnerable, disaster-affected people in Bangladesh." *BMC Public Health* 14(1): 1–9. www.biomedcentral.com/1471-2458/708, accessed on 17/12/2014.

Parvin, G. A., Takahashi, F. and Shaw, R. (2008). "Coastal hazards and community-coping methods in Bangladesh." *Journal of Coastal Conservation* 12(4): 181–193.

Paul, S. K., Paul, B. K. and Routrary, J. K. (2012). "Post-Cyclone Sidr nutritional status of women and children in coastal Bangladesh: An empirical study." *Natural Hazards* 64(1): 19–36.

Paxson, C., Fussell, E., Rhodes, J. and Waters, M. (2012). "Five years later: Recovery from post traumatic stress and psychological distress among low-income mothers affected by Hurricane Katrina." *Social Science & Medicine* 74(2): 150–157.

Phalkey, R., Dash, S. R., Mukhopadhyay, A., Runge-Ranzinger, S. and Marx, M. (2012). "Prepared to react? Assessing the functional capacity of the primary health care system in rural Orissa, India to respond to the devastating flood of September 2008." *Global Health Action* 5, doi:10.3402/gha.v5i0.10964

Phalkey, R., Reinhardt, J. D. and Marx, M. (2011). "Injury epidemiology after the 2001 Gujarat earthquake in India: A retrospective analysis of injuries treated at a rural hospital in the Kutch district immediately after the disaster." *Global Health Action* 4: 7196. doi:10.3402/gha.V4i0.7196

Rahmatullah, M., Ferdausi, D., Mollick, M. A. H., Jahan, R., Chowdhury, M. H. and Haque, W. M. (2010). "A survey of medicinal plants used by Kavirajes of Chalna area, Khulna district, Bangladesh." *African Journal of Traditional, Complementary and Alternative Medicines* 7(2): 91–97.

Rathore, F. A., Gosney, J. E., Reinhardt, J. D., Haig, A. J., Li, J. and DeLisa, J. A. (2012). "Medical rehabilitation after natural disasters: Why, when, and how?" *Archives of Physical Medicine and Rehabilitation* 93(10): 1875–1881.

Ray-Bennett, N. S., Collins, A., Bhuiya, A., Edgeworth, R., Nahar, P. and Alamgir, F. (2010). "Exploring the meaning of health security for disaster resilience through people's perspectives in Bangladesh." *Health & Place* 16(3): 581–589.

Sarwar, M. G. M. (2005). "Impacts of sea level rise on the coastal zone of Bangladesh." Sweden, Lund University, Master's Thesis.

Story, W. T., Burgard, S. A., Lori, J. R., Taleb, F., Ali, N.A. and Hoque, D. E. (2012). "Husbands' involvement in delivery care utilization in rural Bangladesh: A qualitative study." *BMC Pregnancy and Childbirth* 12(1): 28.

Stratton, S. J. and Tyler, R. D. (2006). "Characteristics of medical surge capacity demand for sudden-impact disasters." *Academic Emergency Medicine* 13(11): 1193–1197.

Takasaki, Y. (2011). "Targeting cyclone relief within the village: Kinship, sharing, and capture." *Economic Development and Cultural Change* 59(2): 387–416.

Tunstall, S., Tapsell, S., Green, C., Floyd, P. and George, C. (2006). "The health effects of flooding: Social research results from England and Wales." *Journal of Water and Health* 4: 365–380.

Uddin, J. and Mazur, R. E. (2014). "Socioeconomic factors differentiating healthcare utilization of cyclone survivors in rural Bangladesh: A case study of cyclone Sidr." *Health Policy and Planning* 30(6): 782–790, doi:10.1093/heapol/czu057

WHO. (2006). "Working together for health: The world health report 2006." World Health Organization.

6 Gender and healthcare access after disasters

6.1 Introduction

A gendered analysis of women's healthcare utilization and its influencing factors is a primary focus of this chapter. The previous chapter reveals that environmental and behavioural factors have direct impacts on healthcare access after disasters. Through its qualitative approach, the current chapter demonstrates how environmental, economic, socio-cultural, behavioural factors work together in combination to create great difficulty for women to receive healthcare during and after disasters.

6.2 Gender and impacts of disasters on healthcare access

Disasters have a huge negative impact on accessibility to healthcare facilities for all victims but are intensified for women (Rashid and Michaud 2000; Alam and Rahman 2014) due to factors rooted in unequal power relations between the genders, exaggerated by the social, political and economic subordination of women (Wiest et al. 1994 cited in Fisher 2010). Despite increased health problems and injuries, as Chapter 3 demonstrated, 'gender' is found to override other factors in seeking healthcare irrespective of society or country (Pitmman 1999, Sen 2001; Cannon 2002; Momsen 2010; Karim et al. 2007), and one result is women's lesser access to healthcare after disasters (Alam and Rahman 2014). Rokeya, an aged respondent, reported in her interview, 'Women get more injuries but receive less healthcare facilities than men after cyclones' (Female, age 55, rural area). She has experienced several cyclones, from the great Bhola cyclone in 1970 to Mahasen in 2013. Her description represents women's vulnerability in the remote coastal district of Bangladesh recreated after every cyclone (Alam and Rahman 2014). Several factors influence the healthcare accessibility, as, after Sidr, lack of healthcare facilities, emergency services, disrupted transport and solvency affected most of the victims, but women's constraints became more exaggerated by the socio-cultural factors, such as lack of independence, conservative attitudes, socio-cultural norms, attitude and behavior, which have also been mentioned in several publications (Rashid and Michaud 2000; Neumayer and Plümper 2007; Alam and

Rahman 2014). Jasim compared the problems faced by female victims with those of men and boys in Barguna,

> Both males and females faced problems of healthcare access after disasters . . . Men were vulnerable where they lived far away . . . as the roads were disrupted and they could not come to the municipality, could not visit proper doctors and some faced lack of money, whereas women faced all of these situations and more . . . as society does not pay attention to them like men. Women remain far behind.
>
> (Jasim, age 20, urban area)

To elaborate on Jasim's opinion, if we compare the healthcare accessibility among respondents, no female injured or sick respondent could receive any treatment within 72 hours of the injury, except one woman. Comparatively, men were in better position to collect medicines or have checkups for their injuries or sickness. Some of them could have collected medicines for their injuries even the next day after Cyclone Sidr. Women became highly dependent on their family members to receive any healthcare, as there were no special healthcare facilities for sick/injured, pregnant and postpartum women in the affected study areas. But this dependency lowered women's access to healthcare, whereas the reasons for the men's delayed treatment were found to be the personal approach to injuries, unavailability of transport and economic support.

Dependency and women's healthcare

'Women had to depend on men to get treatment for their injuries and sickness after disasters' was the most common reason respondents gave for their delayed treatments after cyclones Sidr and Mahasen. Dependency on men includes collecting medicines from the distant healthcare centres through inaccessible transport, as there were no emergency services to rescue and transfer patients and no special healthcare services for women on the relief and medical teams, and women needed men's decisions and company to travel to the healthcare centre. But men were mostly busy for relief collections and repairing damage after the cyclones and became unavailable for taking decisions and giving company, which made the women's healthcare access delayed. Here it should be mentioned that delay in decisions and women's healthcare access is common in Bangladesh. Compared with men, women experience longer delays at various stages of healthcare (Karim et al. 2007), and their health and healthcare access are strongly influenced by different socio-cultural factors, such as social norms and role behaviour (Paolisso and Leslie 1995; Enarson and Morrow 1998; Narayan et al. 2000 cited in Fikree and Pasha 2004). During disasters, pre-existing gender discrimination and harassment increase (Rashid and Michaud 2000; Neumayer and Plümper 2007; Alam and Rahman 2014), especially when resources become scarce and unavailable. Unavailability of local healthcare and insufficient relief support increased dependency and intensified the health problems of female victims when they required special health care.

Women's special needs and absence of emergency services

There is a significant difference in the health needs of men and women (Doyal 2000); even the 'need for medicines is gendered' (Momsen 2010, p. 79). Women need special and more healthcare (Momsen 2010; Alam and Rahman 2014) because of their 'reproductive health' and 'women's health' issues (Paolisso and Leslie 1995). During disasters, besides there being a higher number of injuries and deaths among women (Neumayer and Plümper 2007; Alam and Collins 2010; Rahman 2013), they also become vulnerable to reproductive and sexual health problems due to a lack of healthcare facilities (Rahman 2013). Asma explained clearly the special health needs of women: 'Special arrangements are needed for women . . . They have menstrual cycles which might become difficult in disasters . . . Any woman may be a new mother or give birth at that time . . . So, both mothers and babies need special care . . . But these requirements are not necessary for men . . . they do not have any such problems' (Female, age 22, urban area). But female respondents in Barguna did not receive sufficient support from the government and NGOs during the post-disaster periods (Alam and Rahman 2014). After cyclones Sidr and Mahasen, there was no special attention or healthcare provisions for women in Barguna. There were no emergency services to rescue injured women and transfer them to the distant healthcare, no gynaecological doctor/female doctor in the medical team and even the relief packages lacked clothes or sanitary items. The neglect of gender results in relief programmes that do not meet women's needs and increase gendered inequalities (Fisher 2010).

There were many pregnant women and new mothers in the affected areas who needed special attention, but relief teams reached the affected areas without any gynaecological doctor with them. The pre-existing, deficient local maternal health service also became unavailable after the cyclones. Saila could not receive her post-operative treatment, as her doctor could not visit Barguna after Cyclone Mahasen, which increased her health problems (Case Study 6.1 elaborates this point). Pregnant women had to reach distant healthcare centres, which was impossible for most of them (Alam and Rahman 2014). Alam and Rahman (2014) mentioned in their research that pregnant victims of the southern coastal region of Bangladesh had to walk long distances on muddy roads during Cyclone Sidr and give birth in the shelters without any necessary arrangement or medical care. In this present research, Saila's sister, Sumsun, and Saila were at the postpartum stage during cyclones Sidr and Mahasen. They faced several health problems as new mothers without proper healthcare services.

Preference for female medical staff or female volunteers is strong among the respondents (Rashid and Michaud 2000). But after the cyclones, fewer female medical staff were available in Barguna healthcare centres. A female doctor mentioned in her interview that this was due to a lack of vehicles and damaged roads, and she could therefore not attend to her duties for the first few days after Sidr. The only Government Hospital of Barguna also did not have any female doctor (information collected from the Barguna Civil Surgeon Office 2014). An observation in this regard will be helpful to describe the situation. During my stay in Barguna, a young girl was suffering from huge summer boils, and she had to get leave from her school to take rest. She was unable to move properly and suffered for days but was not taken to the doctors, despite living only a ten-minute walk from the doctor's practice. Her mother explained the reason: 'I do not want to take my young girl to a male doctor . . . I kept her at home . . .' (Female, age 38, urban area). Preference for a female doctor was also found among the injured victims of Cyclone Sidr. Rokeya, an aged woman from Tentulbaria village, explained proudly, 'There are some pious women who will die rather visit a male doctor' (Female, age 65, rural area), and she was obsessive about women's conservativeness about male doctors. A similar preference was also mentioned by the aged female respondents in Biswas's research (Biswas et al. 2006). Men also prefer to take their female relatives to female doctors: 'If female doctors were available, I would be very happy to take my mother to them' (Male, age 25, urban area).

Female victims also did not raise their problems with the male relief volunteers (there were no female volunteers in Barguna) (Alam and Rahman 2014). As Mohsena explained, 'Many male volunteers came . . . but the women could not tell health problems to them; if there was any female volunteer that would be very helpful for them' (Female, age 20, urban area). This attitude became a constraint for women after Cyclone Sidr to receive health care. There were similar problems after a major Turkish earthquake, where female survivors felt unable to freely discuss their needs with male relief workers (Enarson 2000), and the efficiency of relief distribution increased when female workers were included (Enarson 2000; Begum 1993).

Most relief packages did not include any feminine items that considered these socio-cultural norms (Begum 1993; Enarson 2000), and this limited opportunities for female victims (Enarson 2000 in Neumayer and Plümper 2007). Due to unfamiliarity, distributed sanitary pads did not meet their needs. 'They rejected those items', as respondents mentioned, and many menstruating women went unclean for days: 'There were no napkins, no clothes . . . women had to be dirty that time, where could they get any help?' (Maksuda, female, age 38, urban area). Sometimes menstruating young girls even had to hide them, standing in water to avoid being seen by anyone, which is very unexpected in this society. Similar conditions of young women were revealed in Rashid's research, which focuses on the flood victims in Bangladesh (Rashid and Michaud 2000). She elaborately explained the difficulties young women faced due to lack of facilities and 'social taboos' associated with menstruation.

Here it should be mentioned that women's needs for their reproductive health differs with cultural traditions. Relief items should be included considering the local cultural context to lessen the disaster effects on reproductive health. Macdonald reported that the 2004 tsunami-affected pregnant victims wanted herbs and remedies for their childbirth which were not available in the relief camp

(MacDonald 2005). Postpartum women also faced a lack of supplies for their infants, such as diapers, similar to the situation in New Orleans after Hurricane Katrina (Zilversmit et al. 2014).

When all these special healthcare needs could not be covered by the relief programmes and local healthcare centres, female victims became dependent on the distant healthcare centres as well as on the male members of the family. Men could reach the distant healthcare centres by walking a long distance or swimming through the congested storm water, which seemed impossible for women. As Badar asked in his interview, 'How can a woman go outside the home after a disaster? A man can jump over the fallen trees but woman cannot . . . A man can swim to cross the water' (Male, age 45, urban area). Again, Liza, schoolteacher of the Tentulbaria school, added in her interview, 'The transport system was so bad that it was not possible to take female patients to the healthcare centres' (Female, age 35, rural area).

Here the question arises: 'are all women physically or mentally unfit for travelling along damaged roads or are cultural practices hindering them?' The answer could be 'cultural practices'. If culturally permitted and safe, women would have dared to travel to collect their own medicines rather than suffering at home. There were many women in this present research who were found to have taken very brave steps to save children or help other victims during the disaster. Hadley et al. (2010) revealed in their research that higher autonomy associated with greater agreement with male members has positive health outcomes among the Uzbekistani women (Hadley et al. 2010). Bruson et al. (2009) revealed in their research that women's autonomy is positively associated with children's health in Northern Kenya (Bruson et al. 2009). Similarly, present research revealed an inverse relation with women's healthcare access after cyclones due to the cultural practice of the 'patriarchal dividend' (Connell 2002, p. 142) and a lack of women's control over decisions in the remote coastal regions of Bangladesh.

Case study 6.1

Saila is a young mother of two children. She is a full-time housewife but continuing her studies at home. She is a student of the degree programme of the Open University of Bangladesh. She was an assistant in the government office but had to leave her job because of the youngest child. Being an earning woman, the wife of a rich man and daughter of a well-off, educated father, she belongs to the rich socioeconomic group of Barguna District. Her family owns a house, an established business and farming land, and she owns money and gold as her own property. I met Saila while visiting another respondent's house near to her house. From the very beginning, she was eager to talk with me and know more about my research. I had to visit that respondent's house several times, and every time, I found her holding her four-month-old son. Being an educated and ex-government staff, she is very interested to meet with professionals, especially female. Every person has different stories in their lives, but some stories are really surprising and unbelievable. Meeting Saila, I found her as a young, very friendly married woman with a smiling face, but her stories made me astonished and think of the gender difference in the society in another way.

Saila and her parents were cheated during her marriage in regards to her husband's education. They were told that he had a Higher School Certificate (HSC), and at that time, Saila herself was studying for a Secondary School Certificate. Being insecure and disturbed by the local boys in her own neighbourhood, her father was in hurry to arrange a marriage for her and did not have the time to authenticate the education of the bridegroom. After the marriage, they found out he had cheated. To avoid the further humiliation and dishonour, Saila and her family did not take any further step to get justice. She is still continuing with her marriage, but she became very determined to complete her study and find a good job and have her own identity. Her husband prohibited her continuing with her studies. As Saila mentioned in the interview, 'He stopped my study . . . he did not let me sit for the examination after the marriage . . . People told him that your wife would leave you if she became more educated than you . . . But I managed to study on my own after 3/4 years of the marriage . . . I want to complete my study and get a good job . . .'. Saila's inspiring story showed her determination and courage. But during my fieldwork, I observed that many women in Barguna could not continue after marriage with their studies. It is assumed as normal that women will stop studying, as they are not expected to earn money after their marriage.

Saila usually visits private chambers, private clinics for normal sicknesses, but in emergencies, she depends on the Government Hospital and Barisal healthcare centres. Recently, Saila faced a bitter and pathetic experience during the birth of her youngest child in 2013. Due to the lack of gynaecological doctors, Saila selected a popular doctor from Barisal who visits Barguna once a week. On her caesarean day, she was in the queue until midnight waiting for the operation. Saila reported that doctor did several operations in a row and was exhausted when he did hers. As a result, the stitches were not properly done. She needed an emergency checkup the following week. Unfortunately, the doctor did not visit that week due to Cyclone Mahasen, and Saila could not get her medicine. As a result, the situation worsened, and she became seriously ill. Local doctors could not manage her condition and complete the treatment. Saila described in her interview: 'Seeing me, doctor became very annoyed . . . he was saying, "I cannot give you more strong medicine, it will have a bad effect" . . . but I was suffering badly . . . I became very upset'. Saila needed to visit the Barisal healthcare centres. These treatments cost a lot of money, dependency on family and suffering for her. After a long treatment, Saila is a little better now but still has not returned to her normal household duties. She is forbidden to do heavy household chores. Often, she feels unwell and has to take a proper rest.

Cyclone Mahasen also brought serious sufferings for Saila in another way. Cyclone warnings were issued just seven days after her caesarean. The nearest government office is used as the cyclone shelter locally, and, according to Saila, a large number of people from the nearby slum usually took shelter with cattle, ducks and hens. The environment of the shelter became overcrowded and absolutely not suitable for a caesarean patient. There was no special arrangement for her in the shelter. She would have had to stand like everyone else for the whole time of the cyclone. Saila dared not to take her newborn child to that shelter and felt insecure herself. She stayed in her tin-made house for the whole time of the cyclone in a state of panic. She described her situation, 'I was so scared for my newborn child . . . I thought storm water will take us away'. Luckily, storm water did not enter their house that time, and they did not need to take shelter at the nearby government office.

When I asked Saila for her suggestions on improving the healthcare conditions of Barguna, she emphasized the need for good medical equipment and good, caring doctors. She also added that more attention and special arrangements should be made for vulnerable groups like pregnant women, people with disabilities, old persons and all women to keep them safe during the disaster.

Saila's experiences represent the overall situation of women in Barguna. Women's rights are violated in their lives in different ways, but many times, these are not mentioned and emphasized due to the social trends and culture. Women become more vulnerable during the disasters because of the health problems and lack of special arrangements according to their needs. Even good economic capacity does not guarantee the inhabitants of Barguna will be saved from suffering. The lack of local healthcare services during cyclone disasters brings suffering to all of the inhabitants in different ways, though maybe with different intensities.

Lack of autonomy in decision-making and chaperoning

Cultural practices of the 'patriarchal dividend' (Connell 2002, p. 142) benefit men as a group and exploit and discredit women and girls and make them vulnerable to abuse and attack. Women's autonomy, 'the ability to make decisions on one's own, to control one's own body, and to determine how resources will be used, without needing to consult with or ask permission from another person' (Brunson et al. 2009, p. 3), is positively associated with health and healthcare access (Bloom et al. 2001 cited in Hadley et al. 2010). In Barguna, respondents mentioned that the practice of the patriarchal dividend and the absence of women's autonomy (Brunson et al. 2009) have become one of the constraints for seeking healthcare after disasters (Enarson and Morrow 1998 cited in Neumayer and Plümper 2007). 'Men' are considered as the main decision-maker, rescuer and carer of the family, whereas 'women' are expected to be 'dependent' members of the society. This cultural practice increased women's dependency and intensified their problems in seeking healthcare, along with the damaged transport system and absence of the emergency services during the cyclones.

Women's dependency on a husband or elder members of the family for seeking healthcare is a common cultural practice in South Asian countries like India, Pakistan, Nepal (Fikree and Pasha 2004; Story et al. 2012) and Bangladesh. Azahar mentioned in his interview, 'It is our culture that father, brothers will take care of their female members of the family' (Male, age 38, urban area).

It is socially expected and accepted that women's health problems will be taken care of by the male members of the family and they will take decisions in regards of their healthcare access, like most other decisions (Story et al. 2012; Alam and Rahman 2014). But difficulties arise when men take decisions without properly knowing the problems and special healthcare needs of women. During the interviews, when men were asked a general question in the interviews, such as, 'Do you think women face more healthcare problems in disasters?', most of the male respondents were less interested to talk about this, and their answers were very short, mentioning only a few points. Maintaining men's identity and 'not [wanting to be] involved in women's matters' are two of the main reasons behind this attitude, and male respondents also mentioned that they were not well informed about women's health in the family (Furuta and Salway 2006; Story et al. 2012).

Conservative attitudes towards male members are common among the female residents of Barguna. Women feel 'shy' or 'dislike' or are 'scared' to tell men about their health problems, even their own male relatives (Karim et al. 2007). Most of the time, information on women's sickness remains within the female members of the family, and male members are informed only at a critical stage. Elder women, especially mothers-in-law or mothers, take decisions whether to tell the male members or not (Furuta and Salway 2006; Parkhurst et al. 2006), whereas most women of Barguna are in a vicious cycle of a lack of education and health information. School attendance of female students drops to only 31.24%, half of the entry percentages, after age 14 (BBS 2006). Again, older women sometimes participate in the 'patriarchal dividend'. According to Connell, when women get dividends from the gendered accumulation process and receive benefits directly from other women, getting help in completing domestic household works, they practice the 'patriarchal dividend' (Connell 2002). The older women often mentioned that as they have maintained their lives with all these constraints, and young woman should also do the same and 'be more patient about health problems'. So, information of women's health problems first needs approval from the older women to reach the men, the main decision-makers of the family. As a result, permission gets delayed in many cases, and the health conditions become worse. Nazem explained the situation: 'It is seen that a young girl will mostly inform her mother who may not always be capable of understanding her problems . . . and mothers sometimes do not speak to the male members . . . and as a result the sickness becomes worse. . .' (Male, age 28, urban area). Killewo et al. 2006 also mentioned in his research that a delayed decision to seek care is a common and significant determinant for delayed treatment during the life-threatening obstetric complications of pregnant women and for maternal mortality in rural Bangladesh.

After cyclones, like the other times, decisions on women's healthcare accessibility depends on the other family members, dominantly on the male members. Men, whatever health knowledge and interest they possess, take decisions according to their availability. They decide when and where women will receive medical attention (Story et al. 2012). Decisions taken on women's healthcare are mostly biased, influenced by the 'son preference' or 'male priority' and economic condition of the family. Some of the following quotations make this point.

> Society does not pay attention to women . . . Still people of our society love to have a son more than a girl child . . . women remain far behind. . .
>
> (Male, age 20, urban area)

> Families pays less attention to women's need for healthcare.
>
> (Male, age 20, urban area)

> Women are neglected in their own home . . . people want to spend more money on sons. . .
>
> (Male, age 48, rural area)

In South Asia, it is a common practice that families pay more attention to son's than a daughter's needs, and this disparity continues through the different stages of women's lives from their childhood (Fikree and Pasha 2004). 'Sons are perceived to have economic, social or religious utility; daughters are often felt to be an economic liability because of the dowry system' (Arnold et al. 1998 cited in Fikree and Pasha 2004, p. 823). This 'son priority' was mentioned by both male and female Barguna residents. This gender inequality leads to systematic devaluation and neglect of women's health and healthcare (Fikree and Pasha 2004).

Rima's experience will be very helpful to explain the point. Rima, resident of Tentulbaria village, was severely injured during Cyclone Sidr and was rescued in a life-threatening condition the morning after. She was kept without any medical treatment for about 20 hours until her husband arrived to take responsibility, despite her father and brothers being present at the site. Nazem explained the tradition in his interview. 'For married women, husbands take the decisions and this becomes constraints for women to receive timely treatment in disasters' (Male, age 28, urban area). Nazem's quotation represents most of the married women in this remote society.

Besides waiting for decisions, women need to be chaperoned to visit healthcare centres, especially after disasters. During the normal daytime, mostly women in Barguna, as well in other coastal regions of Bangladesh, can visit healthcare centres with a chaperon (Alam and Rahman 2014). Other female relatives or older children could be the chaperone in the case of the nearest healthcare centres, but a male family member is always preferable, especially for distant locations. This cultural tradition is also practised in other countries like Nigeria, where women are not permitted to travel far for treatment without being accompanied (Stock 1983 cited in Paolisso and Leslie 1995). Even in the Middle East and Latin America, women cannot travel without their husband's or father's permission (Momsen 2010). But after disasters, when men are mostly busy collecting relief and repairing damage and are therefore unavailable, this cultural practice constrains women's opportunity to visit distant healthcare centres followed by post-disaster insecurity. During Cyclone Sidr, many women remained untreated for days due to unavailability of male chaperones. This difficult situation created a negative impact on injured women which is clearly explained in the following statement by Azhar, an NGO officer who worked with Cyclone Sidr victims, 'They [female victim]) remained untreated for days after the disaster as they needed male company to visit doctors. Everyone [male members] remained busy with relief collections, local healthcare centres were closed . . . women faced more suffering staying at home . . . they cannot go anywhere' (male, age 38, urban area). Even this situation became worse for the 'many women [who] lost their husbands . . . No-one was there to take them to the healthcare centre' (Female, age 38, urban area).

Women and their personal behaviours

Personal behaviours are certainly one key factor determining health status and healthcare accessibility (Paolisso and Leslie 1995). The present research revealed differences between men and women in the way they coped with the disaster impacts (Fisher 2010) and considered their injuries and sickness.

Delaying treatment is found more among the female victims, who are less likely to seek appropriate and early care for disease (Fikree and Pasha 2004). Women were found to hide or neglect their injuries and sickness due to cultural conservativeness, consideration of the cost of treatment and increased responsibilities at home after disasters. Mohsena reported that 'Women usually do not talk about their sicknesses . . . they feel scared or feel shy' (Female, age 20, urban area). Women's shyness or conservativeness is taught from their childhood as a cultural practice in Bangladesh. Rashid and Michaud (2000) mentioned in their research that '[i]n Bangladeshi society, young girls are taught at a very early age to be modest and feel ashamed of their own bodies' (Rashid and Michaud 2000, p. 61).

They are taught to be patient about their health problems; for instance, there is criticism of those who ask for formal assistance during child delivery (Parkhurst et al. 2006). Besides, a family's unequal attitude towards 'girls' from childhood make them feel inferior, subordinate in position (Fikree and Pasha 2004; Alam and Rahman 2014) and conservative. Women only tell about their health problems when they become intolerable.

Again, when health problems of female members are discussed, sometimes they are overlooked or trivialized by the family (Pitmman 1999). Trivialization of women's need for healthcare was also reported by Argentinean women in Pittman's research (1999). He found that both healthcare providers and husbands undervalued women's needs. In other research, Fisher also reported misconduct and poor practices by policemen and hospital staff to Sri Lankan female violence victims, making them feel responsible for the abuse (Fisher 2010).

All these pre-existing cultural practices made female victims feel trivialized after disasters in Barguna. They tried to keep their health problems to themselves and tolerated them until they become critical. Mentioning the women's situation, Asma reported, in a frustrated voice, 'No women could visit the healthcare from our locality the day after Sidr . . . They tolerated until it became intolerable' (Female, age 22, urban area). Asma lives in Barguna municipality near the Government Hospital. Her opinion shows that a location near the healthcare centre did not create a favourable situation for women to collect their injury medicine.

Women also consider several things like treatment costs before they take a decision to inform others or visit a healthcare centre. Nazem described his own experiences: 'Women hide their sickness because they feel shy and again, they think, Allah will help them to be cured . . . they think, visiting the doctors will cost money . . . but this delay brings suffering. . .' (Male, age 28, urban area). Most women in Barguna are dependent on family income for their treatment costs. Women who earn, especially in a poor family, contribute all of their money to the family budget, and there may be no surplus for their own treatment (Paolisso and Leslie 1995). A scarcity of money makes women think twice before getting any proper treatment. They wait to keep the treatment cost low and depend on indigenous or primary treatments (Paolisso and Leslie 1995). Akmal from Tentulbaria described: 'Women do not usually visit the healthcare centres . . . they know their problems . . . but due to lack of money they take primary treatment . . . and when it becomes an emergency, they visit doctors' (Male, age 19, rural area).

Again, the repetitive nature of women's work on a tight schedule to meet the daily needs of the household has a negative impact on women's healthcare accessibility (Paolisso and Leslie 1995), which becomes highly intensified after disasters. Asma mentioned in her interview that 'women could not visit the healthcare centre after cyclone Sidr . . . I myself saw that . . . they were busy repairing the damage at home . . . everything was ruined' (Female, age 22, urban area). Women were very busy making meals for the family after the disaster. Being busy, in most cases, women avoided visiting the healthcare centres for the minor injuries after Cyclone Sidr. They only relied on indigenous cures or antiseptics available at home. Dominelli mentioned in her research on 2004 tsunami victims in Sri Lanka that 'women played key roles in keeping the family together and ensuring that its needs were met, even if theirs were not' (Dominelli 2013, pp. 77–93).

The above discussion helps us to conclude that even women's personal approaches to health and healthcare are found to be determined by complex interconnected factors like culture, traditions and the gendered attitudes of society.

Women's dependency and afterwards

Women's dependency did bring negative impacts on the health of female victims in Barguna. The only woman who was in a critical, life-threatening condition could have received treatment, but she wasn't taken by her husband to a local healthcare centre until 20 hours after the incident. All other female respondents with major or minor injuries remained untreated at home and had to manage on their own (Rashid and Michaud 2000). Compared to the female victims, some injured men collected medicine for themselves, or their family members collected medicine for them, even for minor injuries.

Case studies of a male and a female member of a family from Barguna municipality will help to elaborate these points. Both victims of the family were cut in the hand and leg by flying or rolling objects while taking shelter. Only the man received any medicine, as he went himself to the pharmacy the next day, whereas the female, a young girl, was treated with betel leaf. A question may arise if the level of injuries influenced the difference, but an overall comparison shows that women and girls received less attention than the men and boys. In another case, a woman cut her hand while saving the roof of her house in disaster attack but received no treatment. The reason the male member of the family gave was that 'it was impossible to take her to the healthcare centre' (Male, age 27, urban area). This lady used available antiseptics at home. Again, Saila told the story of her sister, who was a new mother and got injured when running for shelter. She was kept at home without any treatment for days. Though they live in the municipality, her family did not take any emergency measures until transport got better. Sima was worried for her sister and described her helplessness: 'She got injured . . . the stitches [from a caesarean operation] were damaged . . . she could not visit doctors at the bazaar . . . there was no normal transport available . . . they did not take her . . . she had to lie on her bed and suffer . . . I advised her to get treatment as soon as possible as this health conditions might get worse' (Female, age 28, urban area).

Men, gender roles and healthcare access

Compared to the women, the men are in favourable position to receive injury treatment after cyclones. Because of what Connell (2002) describes as the 'patriarchal dividend', men are able to take their own decisions and do not need chaperones to travel. If they can find transport and have the money and time for the treatment, they can receive it earlier than the women. Kamal, a schoolteacher from Tentulbaria village, mentioned his advantage compared to the women of

Barguna: 'men can get the treatment first, then women. . .' (Male, age 48, rural area). Again, men do not hide their sickness or injuries. On the contrary, men's sickness or injury gets the attention of all family members, friends and the health-care providers (Pitmman 1999). Fulon, a female earning respondent, expressed a priority for her husband as the main earning member of the family: 'I think that my husband should stay safe and healthy. If he cannot work how can I get any food?' (Female, age 45, urban area). Her own income does not make her feel safe. So, men's sickness and injuries get the full attention of the family from childhood. Men are habituated to express all of their health problems and take decisions at an early stage of their sickness.

After cyclones, delayed treatment is noticed, even among the male victims. Mostly men were delayed in treatment when they were not sure of the seriousness of the injury and were busy collecting relief and removing debris and clearing the roads. A civil surgeon mentioned that many men's injuries became complicated because they came late to the healthcare centre. Some thought of themselves as being a 'hard man' (Doyal 2000, p. 935) and that 'they could handle their ill-ness on their own' (Pitmman 1999, p. 402). Stereotyping their role as the 'main breadwinner' and maintaining their 'male identity' put men at greater risk (Doyal 2000). They took risks and continued with the duties that are socially expected of them (Hart 1989, Waldron 1995 cited in Doyal 2000). As women of all socio-economic groups are not socially allowed to collect relief, men were the main relief collectors. Though men in general benefited from the inequalities of the gender order, there are some instances where they have to pay a considerable price (Connell 2002; Fikree and Pasha 2004). For example, Faruk in Tentulbaria village could not visit healthcare centres for his injury. He mentioned that he was busy with relief collections. His injury has since become a lifelong health problem for him.

6.3 Behavioural factors: 'they wait until health problems become intolerable'

Lack of knowledge about the future impact of an injury and understanding its gravity worked negatively for many victims, both men and women. If they had known that they would be permanently disabled, they might not have waited and might have tried all the opportunities to get proper treatment. Azhar men-tioned in his interview that 'people remained very busy to repair their damaged houses . . . they depend on indigenous methods . . . lack of awareness works in this regard' (Male, age 38, urban area). Inability to judge the gravity of the situation also makes for delays in taking decisions about pregnancy-related morbidities (Killewo et al. 2006). Mostly people wait until their health condi-tion turns into a serious, observable form. If there was bleeding from a cut, they were more serious about these injuries than if it was inside the skin, on bones or in muscles. Three cases of male injuries turned into lifelong suffering because of people's inability to identify the seriousness of the injury followed by a delay in treatment.

Here it should be mentioned that post-disaster difficulties made victims confused and distracted. Nobody wants extended periods of pain, but people want to spend their limited funds for the right purposes during a post-disaster scarcity, and collecting relief food items and cooking for daily survival were therefore prioritized.

If locally free or low-cost healthcare facilities could be arranged, would they have an impact on getting injured victims timely treatment? The answer could be, if free or low-cost healthcare services could be arranged locally after a cyclone, victims would have more options for receiving treatment. This would save them money and time during their busy schedule to collect the relief or repair the damage. Field observation and interview analysis reveal that, at least, primary checkups and treatment should be arranged locally so that major injuries could be identified, making the victims aware if they need to go for further treatment elsewhere. Adamson et al. (2003) mentioned in their research, 'Tackling inequalities in accessing appropriate healthcare will require further investigation, not just patient behaviour, but also the pathways of care between symptom presentation and health interventions' (Adamson et al. 2003, p. 903).

6.4 Socio-economic conditions, poverty and healthcare access

The role of socio-economic status in seeking healthcare is well documented (Ahmed et al. 2005; Biswas et al. 2006; Karim et al. 2006; Karim et al. 2007; Peters et al. 2008; Mahmood et al. 2010; Uddin and Mazur 2015). Poverty makes for delays in healthcare accessibility and influences patients to receive treatment from the low-cost informal sector (Ahmed et al. 2005; Uddin and Mazur 2015). Disaster victims are also influenced by their socio-economic status, as revealed in the present research and supported by other researchers (Uddin and Mazur in 2015). Uddin and Mazur (2015) mentioned that poor socio-economic status influenced Cyclone Sidr victims to rely on self-care or indigenous materials, seek healthcare from drugstore salespeople and village doctors and use other informal care providers (Uddin and Mazur 2015).

Though using different types of indigenous materials is common (Rahmatullah et al. 2010; Uddin and Mazur 2015), it is mandatory for poor victims. They cannot afford to buy antiseptic cream and store it at home in case of an emergency. Paru from the Asrayan (Government residence for homeless people), Barguna municipality, explained the situation, 'We do not have anything at home, we cannot buy, we do not have that economic solvency . . . If injured, we apply powdered ashes, tooth powder or bind with Andow pata (leaves) . . . We cannot do anything else' (Female, age 30, urban area).

Poor victims reported that the day after Sidr, they were helpless, as they could not get proper treatment from the overcrowded hospital or buy medicines from the pharmacies due to lack of money. Most depended on relief items, even though they were not distributed properly in amount or type. Rokeya, an aged poor women (landless) living in Tentulbaria village, described her helplessness after Cyclone

Sidr. 'They gave us medicines . . . not much . . . we are the poor people . . . we had to rely on their supplied medicines whatever they gave us . . . we had to wait . . . we could not visit to doctors . . . we have no money' (Female, age 65, rural area).

Lack of money for treatment after disasters also prevented poor victims from receiving timely healthcare services. Sahida's husband was hurt during Sidr, but he had to wait two to three days before getting proper treatment. She described the situation. 'He started to feel difficulty in hearing after 2–3 days of injury . . . he got hurt while taking shelter during cyclone Sidr . . . a tree fell on him . . . he could not visit any doctor for 2–3 days, where we can go during disasters? After all, some money should be at home . . . I did not have any money that time and could not visit the doctors timely . . . Now he cannot hear well' (Female, age 35, urban area) (details in Case study 3, Box 6.2). To 'have to have money upfront' for the proper timely treatment was also mentioned by the poor respondents in Biswas's research (Biswas et al. 2006, p. 76).

'Neglect of an injury' was also more common among the poor victims in Barguna and in other coastal areas of Bangladesh. Gelberg et al. (2000) mentioned that 'homeless persons are willing to obtain care if they believe it is important' (Gelberg et al. 2000, p. 1273). Again, Parvin mentioned in her research that 'being rural poor, people usually do not seek treatment until the illness becomes severe' (Parvin et al. 2008, p. 188). Minor injuries do not get the proper attention. After a disaster, the collection of food becomes the top priority, and poor victims remain busy collecting the scanty relief items. This delays their treatment and leads them to self-exclusion (Karim et al. 2006; Karim et al. 2014). Doyal mentioned in his article that '[p]overty can prevent both men and women from realizing their potential for health' (Doyal 2000, p. 932), which is also noticeable among the poor victims of Cyclone Sidr in Barguna. Parvin et al. (2008) also revealed the same personal approach among poor cyclone victims in other coastal regions of Bangladesh.

Poverty status has also emerged as a strong determinant of healthcare expenditure (Ahmed et al. 2005) and the types and places of treatment (Biswas et al. 2006). Poor households were found to be avoiding costly allopathic treatment (Biswas et al. 2006) because visiting a formal physician is between five and 15 times as costly compared with non-qualified medical providers (Bhuiya 2009 cited in Uddin and Mazur 2014). The poor mostly chose self-care/self-treatment (home remedies and over-the-counter drugs) and non-qualified medical providers (Ahmed et al. 2005; Uddin and Mazur 2015). Rima, a poor female victim of Cyclone Sidr, was taken to the free government centre for emergency care, but for further treatment, she chose a less expensive, non-qualified medical provider in Barguna municipality. Such non-qualified practitioners are commonly chosen because they are well known to the patients (Biswas et al. 2006; Mahmood et al. 2010) and more likely to accept deferred payment or payment in kind instead of cash (Biswas et al. 2006 cited in Uddin and Mazur 2015). By comparison, better-off victims like Saila and Kamal in the present research received their treatment from allopathic, qualified private doctors from Barguna municipality.

Nazem mentioned in his interview that sometimes, poor, injured people had to borrow money after Cyclone Sidr to get treatment for their injuries, but without a loan, others suffered without treatment: 'many people were living in suffering due to lack of money . . . Some of them [injured persons] asked for money from wealthy persons for their injury treatment' (Male, age 28, urban area). Similar findings have been also mentioned in other research (CCC 2009; Karim et al. 2014).

Again, many injured victims could not complete their treatment because of poor economic conditions and the unavailability of external financial help. There was no relief programme for long-term follow-up treatment provision for injured persons in Barguna. Patients had to seek private practitioners, but many lacked the funds (Paul 2010). Again, most complete injury treatment is not available in Barguna, and patients had to travel outside, for which they needed to find the cost of transport as well as the treatment costs, and there were long hours away from paid work, which was a burden for poor victims (Karim et al. 2006). Rehabilitation could have been helpful, but a lack of rehabilitation resources is common in developing countries (Rathore et al. 2012). Experience shows that victims of spinal cord injuries were in greatest need after the 2005 earthquake in Pakistan because of their disabilities, but they suffered from a lack of rehabilitation resources (Johnson and Galea 2009). Rima, Sahida and Faruk, participants in the present research, were found still to be facing health problems seven years after Sidr, and they described their injuries as 'pain', 'suffering', 'helplessness' and 'burden' during their interviews.

Case study 6.2

Sahida is a middle-aged woman, a temporary housemaid in Barguna municipality. It was the late afternoon when I first met her, and she had just finished her daily duties in her employers' houses. This is the only time she could talk with me because upon returning home, she needed to start her own housework, cook dinner for the family and take care of her five young children. When I saw her, I found a slim woman, wearing a very torn sari with a strong face which bears all the signs of a life of struggle. Though 35, she looked much older and more mature. At first, she seemed a little confused about whether to spend time with me, as she thought I was one of the people who are only interested in survey information and do not listen to experiences and opinions. But when I introduced myself and described my research, she became very eager to talk and share her experiences in cyclones Sidr and Mahasen. Between the two cyclones, it was Sidr that changed her life. Her husband, the main earning member of the family, was injured and became deaf due to a lack of proper treatment. She belongs to a very poor, large family with no property except a house near Barguna municipality. She works at people's houses as a temporary maid and her husband pulls a rickshaw, and both of them earn a very small amount of money for a large family. (It is common for women in Bangladeshi society to work as temporary maids in someone else's house when in extreme poverty.) Sahida needs to work every day from morning to late afternoon to support her family. The total income of her family is 9,000 taka for seven members with children, which is the equivalent of £0.557 per head a day, which does not compare favourably with the extreme poverty line of $1.25 a day per head commonly used as a world development indicator. During any normal sickness, she and her family use the Government Hospital, and, in emergencies, the Government Hospital and private chambers are both visited. Visiting the private practitioners and buying a full course of medicine is very costly and a burden for her.

During Cyclone Sidr, her husband was hit on the head by a fallen tree when they were going to take shelter in a neighbour's house. After the incident, they could not visit the healthcare centre immediately (next day) because they did not have any money to pay the treatment cost. It was also very difficult to move after the disaster. She mentioned that after two to three days, he began to experience hearing difficulties; they went to several private chambers in Barguna municipality and started the medications. But the prescribed medicines did not work, and no improvement was felt. So, they had to visit the healthcare centre of Barisal to find a proper treatment. Unfortunately, the treatment could not bring any improvement, and she was advised to take him to Madras in India. But that is impossible for her, so they left the treatment incomplete because of a lack of money. His hearing has further deteriorated, but they could not even buy a hearing aid six years after the incident. She explained that a 'hearing aid needs lots of money . . . at least 10,000 taka . . . we could not find that amount'. Being very poor, Sahida and her children became vulnerable members of society, as the main breadwinner, her husband, cannot work with his full strength. She repeatedly expressed her helplessness: 'That time I had no money . . . I could not take him to the doctor immediately'. But this poor, injured family did not receive any relief or help after the cyclone. She mentioned with deep sadness, frustration and anger that some other families got relief more than once, whereas they did not get any. She claimed that the families who received the relief several times were in the chain of corruption and mismanagement of the relief distribution.

Sahida and her family's vulnerable situation has not changed six years after the cyclone. In Mahasen in 2013, they were at their same vulnerable house during the warning time, and when they found the situation became more dangerous, they took shelter at the neighbour's house. The main reasons for not taking proper shelter during the warning period was that the warning level was not as high as for Sidr, and they felt insecure leaving the house because of the risk of burglary and theft and also the uncomfortable environment when taking shelter in the neighbour's house. So, they were again in an extremely vulnerable condition. They were again in a race to save their lives.

6.5 Poverty, gender and healthcare access

People living in poverty are much more vulnerable to the effects of natural disasters (MacDonald 2005), and the low socio-economic status of women and their place in social norms and role behaviour make them the most vulnerable victims (Neumayer and Plümper 2007). In developing countries, inadequate medical services, along with financial crises and gender discrimination (Doyal 2000), compound the problems of women and girls in seeking healthcare (Paolisso and Leslie 1995; Chowdhury et al. 2003; Sultana 2010; Story et al. 2012). All of this is intensified after disasters (Neumayer and Plümper 2007; Uddin and Mazur 2015).

When relief was the only opportunity for the poor victims to receive any medical help, the gendered attitudes of the society and family towards women often marginalized their access to relief (Neumayer and Plümper 2007). Similar findings were revealed in Alam and Rahman's research, where 82.86% of female survivors mentioned deprivation of relief material during cyclones (Alam and Rahman 2014). Young girls or young women are not expected to collect relief from the roadside or relief queue after disasters, whereas men of all ages and sometimes older women from the same families are allowed. This attitude of society created a vulnerable situation for poor injured/sick young/middle-aged women after Cyclone Sidr in the present study areas. Women had to remain starving or without medicines until the male members brought the relief aid to them at home. Azahar described the situation in his interview: 'A boy from a middle-class family can collect relief from the roadside whereas his sister cannot . . . women are highly discriminated against after disasters . . . they suffer with hunger and health problems' (Male, age 38, urban area). Rokeya from Tentulbaria village explained, 'My sons collect the relief or old women like me collects the relief . . . then we divide the relief among the children and daughters-in-law . . .' (Female, age 65, rural area). In this family, daughters-in-law are the female members of the family.

Here it should be mentioned that though in many newspapers and journals, photographs were published of queues of women collecting relief after the cyclones, the present research revealed that all women, especially from solvent families, could not collect relief, even in extreme need. Social humiliation was the main reason behind this behaviour. These women had to suffer in need of medicine and food because of the gendered attitude of society during the very difficult post-cyclone conditions. Azahar mentioned this from his experience as an NGO officer: 'There were some poor women with higher social status ... they suffered at home but cannot ask for help from others ... There was an injured woman who remained untreated with a broken arm' (Male, age 38, urban area).

Dependency on an inadequate family income increased poor women's vulnerability to health problems. According to Nazem, a male respondent from Barguna municipality, 'if the family has a shortage of money, the women in the family suffer more in regards to healthcare accessibility' (Male, age 28, urban area). Nazem's opinion could be elaborated with the findings mentioned in other research (CCC 2009), that if men become sick, they borrow money for healthcare, whereas 'women do not even dare to think of affording healthcare, they just share with someone trustworthy and accept any consequences' (CCC 2009, p. 49). Again, Paolisso and Leslie commented that 'in attempting to meet their health needs, women in developing countries, particularly those in low income families, face a number of economic, social and cultural constraints' (Paolisso and Leslie 1995, p. 60). Besides, a distant location from healthcare centres increases the treatment cost for women as they need extra money for transport, which is not always affordable (Chowdhury et al. 2003). Women cannot walk long distances or share public transport, such as the motorcycles in Tentulbaria village. So, poor women keep their problems unspoken or use available remedies at home. In the present, as well as in other research, dependency on available remedies as indigenous materials is more noticeable among the poor female victims (Alam and Rahman 2014; CCC 2009). Asma and Alauddin's mother, Maksuda's niece in Barguna only used betel leaves and antiseptic creams to heal their injuries after Cyclone Sidr.

As we have noted, poor families do not always pay attention to the healthcare needs of women and girls (Doyal 2000) and do not want to spend money on them. In Barguna, Rina and Saila's sister were seriously injured during Cyclone Sidr but did not receive proper treatment. Rima needs further operations for her full recovery, but her economic dependency has created an insurmountable obstacle.

At this point of the discussion, we might expect that education and employment might help women to receive the proper treatments, but the following sub-section shows that even socio-economic and educational statuses do not always assure women's timely access to the healthcare centres due to social attitudes and culture.

6.6 Intersectionality, gender and healthcare access

Intersectionality can be defined as 'the mutually constitutive relations among social identities' (Shields 2008, p. 301). It is 'useful as a handy catchall phrase that aims to make visible the multiple positioning that constitutes everyday life and the power relations that are central to it' (Phoenix and Pattynama 2006, p. 187). Intersectionality has become a central principle in feminist thinking and popular in explaining complex gender relations (Brah and Phoenix 2004; Shields 2008; Phoenix and Pattynama 2006). Intersectionality offers a way of understanding the complex relations of gender identities (masculinity and femininity) with other social variables. As Butler (1990) has contended, it is impossible 'to separate out "gender" from the political and cultural intersections in which it is invariably produced and maintained' (Butler 1990, p. 5). An intersectional analysis was conducted in the present research to reveal the complex relationships between gender and healthcare access during disasters. It was expected that 'all women do not suffer to the same extent or in the same ways' (Byrne and Baden 1995 cited in Fisher 2010, p. 904). However, the analysis of interlocking social and economic identities in Barguna is not straightforward. The findings here are complex, raising further theoretical questions, which are discussed in the following paragraphs.

Intersectional analyses have been conducted considering cross-sectional factors like education, employment, age and location of people's residences. The findings reveal that higher education has a positive impact on the inhabitants; it makes them become very aware of health issues and the quality of healthcare facilities. Especially, higher education profoundly influences women, increasing their health awareness (Chowdhury et al. 2003) and giving them opportunities to participate in the family income as well as decision-making (Furuta and Salway 2006). According to a respondent who has worked on women's health for 21 years as a government employee, 'about 10% of female patients visit male doctors for their reproductive health problems, most of them are educated and economically solvent whereas illiterate and poor women avoid visiting male doctors' (Female, age 48, urban area). Again, during the interviews, I observed that educated and earning women are mostly affluent and confident about their answers, with some exceptions in the poor families. Liza, a female teacher of Tentulbaria village, mentioned that 'woman's own socio-economic status has an influence on the decisions taken in the family whether to visit healthcare centres' (Female, age 35, rural area), which she experienced during her first child delivery operation. She believes that higher education and income status helped her to participate in the decision about having a caesarean. Here it should be mentioned that taking decisions in a patriarchal society is solely considered as a masculine role. However, the positive relation between a woman's socio-economic status and the health impacts of disaster is also mentioned by Neumayer and Plümper in 2007. They found better resilience to the adverse impacts of natural disasters (Alam and Rahman 2014). Saila, an educated female victim, sought out an MBBS doctor rather than visiting available non-qualified medical providers after Cyclone Mahasen. Uddin and Manzur (2015) also revealed in their research that cyclone survivors with a formal education were more likely to seek allopathic treatment than those with no education (Uddin and Mazur 2015). But here it should also be mentioned that the higher awareness of educated women does not assure them timely access to proper healthcare facilities during disasters, a problem they share with poor earning women. They might participate in the family decision; however, the family (male members and, sometimes, senior females) takes the final decision. Social attitudes and culture have a stronger influence on women's access to healthcare than their own socio-economic status and awareness.

Again, during recent years, more women are getting a chance for employment, especially in garment factories and other informal sectors in Bangladesh. But these earnings do not always help poor women to enjoy freedom or autonomy; rather, 'very few have any control over their own earnings' (Furuta and Salway 2006,

p. 17; Alam and Rahman 2014). They may get an opportunity to express their opinions, but the final decisions are taken by their family. Several studies show that poor women in Bangladesh may receive loans or credits or earn money, but their loans or earnings are mostly controlled by their male relatives, especially by their husbands (Goetz and Gupta 1996; Ahmed 2004). Ahmed mentioned that '[m]ost married women are unable to leverage their income into greater decision-making power' (Ahmed 2004, p. 34), whereas women's earnings are essential for their family maintenance, and the demands of the family rarely leave any surplus for their own treatment (Paolisso and Leslie 1995). So, the question arises: 'is society changing in favour of women's freedom or for some other benefits?' If these changes do not bring enough transformation in women's status, they are again being burdened with responsibilities, as Momsen mentioned, the 'triple burdens' of housework, childcare and earnings or food production (Momsen 2010, p. 2). In that case, an increase of awareness of gender equity and equality is essential.

Age is an important factor which has different influences on the residents. Both older men and older women become more vulnerable in disasters, especially when they need support to take shelter or get treatment. There were several cases revealed in the present research where older people were left behind when they were taking shelter in disasters, and they needed help to get food and healthcare access. Their dependency on the family and neighbours increased their vulnerability. Young children also become very vulnerable. However, age in some cases becomes a source of privilege for older women, as they become one of the decision-makers in the family. Nonetheless, problems arise when they become complicit with patriarchy, taking decisions with male family members. This trend is noticeable mainly when older women play the role of mothers-in-law. Some of these older women want to continue with tradition, preferring younger women to be dependent. However, these traditions work negatively during disasters when the number of victims becomes higher than the number of rescuers. Women's dependency creates an adverse impact on their health and healthcare access.

The location of residences also creates differences among the inhabitants. Most of the stronger houses in the locality, healthcare centres, government and private offices, and educational institutes are located in urban areas. These facilities help to increase resilience (they are stronger buildings in safer locations) and higher healthcare access in urban areas compared to remote rural areas. Even so, urban residents, especially urban women, were found to be more aware about health and healthcare facilities than rural women. However, after disasters, strong cultural attitudes, beliefs, norms and traditions create homogeneity among urban and rural women in healthcare access. Both urban and rural women received less healthcare after disasters, despite facing more health problems.

In the present research, utilizing the concept of intersectionality helps us to understand that 'the individual's social identities profoundly influence one's beliefs about and experience of gender' (Shields 2008, p. 301). It reveals the importance of understanding masculinity and femininity in 'the context of power relations embedded in social identities' (Collins 1990; 2000 cited in Shields 2008, p. 301). These social identities create differences among residents. Socio-economic conditions and the location of the residence increase physical protection, resilience and improve awareness of health and healthcare access of the inhabitants but do not increase and assure women's healthcare access after disasters. In a patriarchal society, women are more confined with feminine roles and remain dependent, whereas men are more privileged and take decisions. Here it should be mentioned that there have been recent changes (discussed in the next section) in Barguna society which bring some hope for the future in regards of shifting inequality. Nevertheless, a serious question arises about the future of such social changes in such a patriarchal society where gender identities and roles are strongly followed. For Butler (1990), gender is performative and is reiterated by 'the repeated stylisation of the body, a set of repeated acts within a highly regulated frame that congeal over time to produce the appearance of substance, of a natural sort of being' (Butler 1990, p. 45). In Barguna, many of my respondents (male and female) expressed some desire for change, as discussed in the following section, but at the same time, they often reiterated traditional roles through their practices of masculinity and femininity. Therefore, gender roles are very deeply rooted, and the complex relations of gender with other social variables mean that achieving meaningful social change that will benefit the most vulnerable women is a long-term endeavour.

6.7 Gender and culture: some changes and hopes for the future

> Now people have become more aware of their daughters . . . they take them to doctors at the early stage of the sickness
>
> —(male, age 28, urban area).

Alauddin and few other respondents mentioned recent changes in Barguna society in regards of gender inequality. They reported that there is a low but increasing awareness among the inhabitants of the need to value their daughters, sending them to school and giving them more opportunities to receive basic needs. Sen (2013) also mentioned in his article on Bangladesh that 'one direction of change is the emphasis that the country has placed on reducing gender inequality in some crucially important respects' (Sen 2013, p. 1966).

In Bangladesh, in recent years, people are sending more girls to primary school (Chowdhury et al. 2002), which makes Bangladesh one of the few countries in the world where the presence of girls is higher than boys at that level (Sen 2013). This is a great achievement for this country, but questions arise about the causation of this move. Rashid claims that parents hope that 'their daughters will attract suitable husbands, but also any schooling or training may make the girls more employable' (Rashid and Michaud 2000, p. 55). If so, this might explain the lower presence of female students in the higher secondary level. According to Chowdhury, 'The "real" net enrolment rate was found to be 73% at the primary level, it is only 13 % at the higher secondary level' (Chowdhury et al. 2002, p. 202). Most girls, especially in remote areas, are married at age 15–16 years.

But there is hope. Nowadays, Purdah is no longer a constraint on women to visiting a male doctor in an emergency, even in the remote Tentulbaria village. Women widely use the 'Borka' (a dress which covers the whole body) and cover their heads and highly prefer female doctors, but they are not prohibited from visiting male doctors in an emergency. As Sardar and Runa mentioned in their interviews, 'Purdah is not a constraint for healthcare services' (Male, age 18, rural area), and Runa said, 'There is no restriction for male or female doctors . . . my mother-in-law does not tell anything about it . . . in Barguna city when only male doctors are available we visit them in emergency' (Female, age 30, rural area).

Some people are trying to become more aware and change their attitudes. Nazem's quotation from his interview is a case in point: 'There are still some people who are against the higher education of women . . . who prohibit women from healthcare services . . . I myself was also against the higher education of women . . . But now I meet with educated people and I read newspapers . . . I became aware . . . I got rid of that mentality' (Male, age 28, urban area).

Actually, from my own observations, some men do not want to practice the cultural tradition of gender differences. Momsen mentions in her book that 'even in highly patriarchal societies some men can remain marginal to the dominant order of patriarchy and be open to change' (Momsen 2010, p. 106). Due to the 'patriarchal dividend' (Connell 2002) and due to teasing about their 'manliness' by their friends, some men cannot come forward for gender equality. Besides, 'in a patriarchal society being male is highly valued, and men value their masculinity' (Kaufman 1987, p. 496). The treatment of masculinity and femininity in the context of inter-related social variables increases inequalities among inhabitants. Both men and women reproduce and reinforce gender inequalities, maintaining their gender identities and roles. In the present research, men mentioned their alertness during Sidr for cultural tradition and insecurity. Sending their wife to collect relief may end in their humiliation, like the case study mentioned in Momsen's book. It shows that a man in Tajikistan was taunted and called names by his own father because of his dependency on his wife to maintain the family (Momsen 2010, p. 97).

6.8 Conclusion

Reflecting the major issues raised by interviewees during the study, the last two chapters (Chapters 5 and 6) focused on several factors and their relationships to healthcare utilization by disaster victims. Chapter 5 describes the environmental factors, whereas this chapter focuses on the demographic, behavioural, social, cultural, economic and health factors which are inter-related and interact in a complex way and affect individual's access to healthcare in disasters significantly (Figure 6.1).

During the immediate post-disaster periods, healthcare access is extensively affected by environmental factors (e.g. availability of local healthcare facilities and medical relief aid and accessibility of transport systems), combined with other inter-related factors like socio-cultural factors (e.g. gender and cultural attitudes, availability of family and social supports); economic factors (e.g. ability to

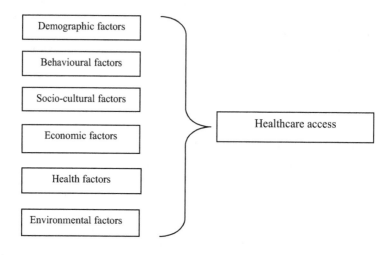

Figure 6.1 Factors affecting healthcare access in disasters

pay for the treatments); health factors (e.g. severity of injury); and behavioural factors (perception of severity of the conditions and health seeking behaviours). All these factors, especially economic conditions, socio-cultural and behavioural factors, affect the time of healthcare access, place of treatment and completeness of the injury treatments, which have been discussed elaborately in this chapter and in Chapter 5.

The present research therefore expands on and fleshes out the indications of these factors in existing literature. Most importantly, through its qualitative approach, the current research demonstrates how these factors work together in combination to create great difficulty for the most marginalized people in this region. For example, Few and Tran 2010 focused on economic livelihood, household location and health awareness or education and their inter-relations as the core factors influencing the health impacts of climatic hazards in hazard-prone Vietnam. Parvin et al. (2008) revealed economic conditions, location and transportation system as significant factors influencing healthcare access in disasters among the

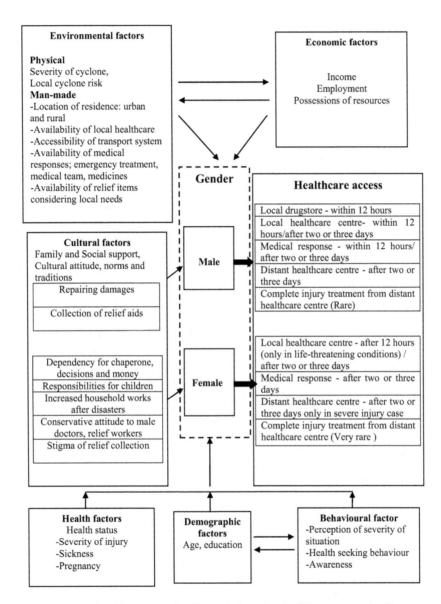

Figure 6.2 Relationships among the factors influencing healthcare access in disasters: a gendered analysis

disaster-prone coastal Islanders in Bangladesh. Uddin and Mazur (2015) in their research revealed the strong influence of socio-economic conditions on healthcare utilization among the Cyclone Sidr survivors, including behavioural factors as perceived susceptibility to post-disaster water-borne diseases linked with health risk communication and education. Boscarino et al. also mention behavioural factors; 'they did not believe they had a problem' (Boscarino at al. 2005, p. 287) was one of the major reasons for not seeking treatment for mental health problems among the inhabitants of New York City after the World Trade Center disaster (Boscarino et al. 2005); this is a significant reason too among the Barguna inhabitants, as described in the previous sections of this chapter. These factors have also been mentioned in other research (Johnson and Galea 2009).

A gendered analysis of women's healthcare utilization and its influencing factors has been a primary focus of this chapter (Figure 6.2). These factors have also been mentioned in other research (Puentes-Markides 1992; Paolisso and Leslie 1995; Anwar et al. 2011).

Analysis of these factors reveals (Figure 6.2) that socio-cultural factors, especially cultural attitudes to gender relations and responsibilities, have a significant influence on healthcare access of women after disasters, being intensified in poor economic conditions. Women are taught to be dependent and treated as dependent family members, which makes them victims of social attitudes, beliefs, norms and culture, as well as the adverse impacts of disasters on healthcare access. Inaccessible transport creates difficulties in reaching medical centres, while women's health status, such as pregnancy, malnutrition and cultural attitudes (women's dependency for decisions, the need for chaperones, as well as lack of money, resources and autonomy) exaggerate their problems in disasters. The influence of these cultural attitudes is so strong that higher education and income do not assure women's healthcare access. The influence of culture has also been mentioned as important in shaping women's healthcare access in other research (Neumayer and Plümper 2007; Alam and Rahman 2014; Nahar et al. 2014; Sultana 2010).

References

Adamson, J., Ben-Shlomo, Y., Chaturvedi, N. and Donovan, J. (2003). "Ethnicity, socio-economic position and gender – do they affect reported health-care seeking behaviour?" *Social Science & Medicine* 57(5): 895–904.

Ahmed, F. E. (2004). "The rise of the Bangladesh garment industry: Globalization, women workers, and voice." *NWSA Journal* 16(2): 34–45.

Ahmed, S. M., Tomson, G., Petzold, M. and Kabir, Z. N. (2005). "Socioeconomic status overrides age and gender in determining health-seeking behaviour in rural Bangladesh." *Bulletin of the World Health Organization* 83(2): 109–117.

Alam, E. and Collins, A. E. (2010). "Cyclone disaster vulnerability and response experiences in coastal Bangladesh." *Disasters* 34(4): 931–954.

Alam, K. and Rahman, M. H. (2014). "Women in natural disasters: A case study from southern coastal region of Bangladesh." *International Journal of Disaster Risk Reduction* 8: 68–82.

Anwar, J., Mpofu, E., Mathews, L., Shadoul, A. F. and Brack, K. E. (2011). "Reproductive health and access to healthcare facilities: Risk factors for depression and anxiety in women with an earthquake experience." *BMC Public Health* 11(1): 523. www.biomedcentral.com/1471-2458/11/523

Arnold, F., Choe, M. K. and Roy, T. K. (1998). "Son preference, the family-building process and child mortality in India." *Population Studies* 52(3): 301–315.

BBS. (2006). *Population Census-2001, Community Series*. Zila, Barguna, Bangladesh Bureau of Statistics.

Begum, R. (1993). "Women in environmental disasters: The 1991 cyclone in Bangladesh." *Gender & Development* 1(1): 34–39.

Bhuiya, A. (2009). *Health for the Rural Masses: Insights From Chakaria*, Bangladesh, ICDDRB.

Biswas, P., Kabir, Z. N., Nilson, J. and Zaman, S. (2006). "Dynamics of health care seeking behaviour of elderly people in rural Bangladesh." *International Journal of Ageing and Later Life* 1(1): 69–89.

Bloom, S. S., Wypij, D. and Das Gupta, M. (2001). "Dimensions of women's autonomy and the influence on maternal health care utilization in a north Indian city." *Demography* 38(1): 67–78.

Boscarino, J. A., Adams, R. E., Stuber, J. and Galea, S. (2005). "Disparities in mental health treatment following the World Trade Center disaster: Implications for mental health care and health services research." *Journal of Traumatic Stress* 18(4): 287–297.

Brah, A. and Phoenix, A. (2004). "Ain't I a woman? Revisiting intersectionality." *Journal of International Women's Studies* 5(3): 75–86.

Brunson, E. K., Shell-Duncan, B. and Steele, M. (2009). "Women's autonomy and its relationship to children's nutrition among the Rendille of northern Kenya." *American Journal of Human Biology* 21(1): 55–64.

Butler, J. (1990). *Gender Trouble: Feminism and the Subversion of Identity*. New York, Routledge.

Byrne, B. and Baden, S. (1995). *Gender, Emergencies and Humanitarian Assistance*. Brighton, Institute of Development Studies.

Cannon, T. (2002). "Gender and climate hazards in Bangladesh." *Gender & Development* 10(2): 45–50.

CCC. (2009). *Climate Change, Gender and Vulnerable Groups in Bangladesh*. Dhaka, Climate Chamge Cell, DoE, MoEF, Component 4b, CDMP, MoFDM: 1–82.

Chowdhury, A. M. R., Bhuiya, A., Mahmud, S., Salam, A. K. M. A. and Karim, F. (2003). "Immunization divide: Who do get vaccinated in Bangladesh?" *Journal of Health, Population and Nutrition* 21(3): 193–204.

Chowdhury, A. M. R., Nath, S. R. and Choudhury, R. K. (2002). "Enrolment at primary level: Gender difference disappears in Bangladesh." *International Journal of Educational Development* 22(2): 191–203.

Collins, P. H. (1990). *Black Feminist Thought: Knowledge, Consciousness, and the Politics of Empowerment*. Boston, Unwin Hyman.

Collins, P. H. (2000). *Black Feminist Thought: Knowledge, Consciousness, and the Politics of Empowerment*, 2nd edition. New York, Routledge.

Connell, R. W. (2002). *Gender*. Cambridge, Polity Press.

Dominelli, L. (2013). "Gendering Climate Change: Implications for Debates, Policies and Practices." In Alston, M. and Whittenbury, K. (eds.) *Research, Action and Policy: Addressing the Gendered Impacts of Climate Change*. London, Springer: 77–93.

Doyal, L. (2000). "Gender equity in health: Debates and dilemmas." *Social Science & Medicine* 51(6): 931–939.

Enarson, E. P. (2000). *Gender and Natural Disasters*. Geneva, ILO.

Enarson, E. P. and Morrow, B. H. (eds.) (1998). *The Gendered Terrain of Disaster*. New York, Praeger.

Fikree, F. F. and Pasha, O. (2004). "Role of gender in health disparity: The South Asian context." *BMJ* 328(7443): 823–826.

Few, R. and Tran, P. G. (2010). "Climatic hazards, health risk and response in Vietnam: Case studies on social dimensions of vulnerability." *Global Environmental Change* 20(3): 529–538.

Fisher, S. (2010). "Violence against women and natural disasters: Findings from post-tsunami Sri Lanka." *Violence Against Women* 16(8): 902–918.

Furuta, M. and Salway, S. (2006). "Women's position within the household as a determinant of maternal health care use in Nepal." *International Family Planning Perspectives* 32(1): 17–27.

Gelberg, L., Andersen, R. M. and Leake, B. D. (2000). "The behavioral model for vulnerable populations: Application to medical care use and outcomes for homeless people." *Health Services Research* 34(6): 1273–1303.

Goetz, A. M. and Gupta, R. S. (1996). "Who takes the credit? Gender, power, and control over loan use in rural credit programs in Bangladesh." *World Development* 24(1): 45–63.

Hadley, C., Brewis, A. and Pike, I. (2010). "Does less autonomy erode women's health? Yes. no. maybe." *American Journal of Human Biology* 22(1): 103–110.

Hart, N. (1989). "Sex, gender and survival: inequalities of life chances between European men and women." in J. Fox (ed.) *Health inequalities in European countries*. Aldershot, UK, Gower Publishing Limited, 109–141.

Johnson, J. and Galea, S. (2009). "Disasters and Population Health." In K.E. Cherry (ed.) *Lifespan Perspectives on Natural Disasters*, London, Springer: 281–326.

Karim, F., Islam, M. A., Chowdhury, A., Johansson, E. and Diwan, V. K. (2007). "Gender differences in delays in diagnosis and treatment of tuberculosis." *Health Policy and Planning* 22(5): 329–334.

Karim, M., Castine, S., Brooks, A., Beare, D., Beveridge, M. and Phillips, M. (2014). "Asset or liability? Aquaculture in a natural disaster prone area." *Ocean & Coastal Management* 96: 188–197.

Karim, F., Tripura, A., Gani, M. S. and Chowdhury, A. M. R. (2006). "Poverty status and health equity: Evidence from rural Bangladesh." *Public Health* 120(3): 193–205.

Kaufman, M. (1987). "The Construction of Masculinity and the Triad of Men's Violence." In M. Kaufman (ed.) *Beyond Patriarchy: Essays on Pleasure, Power and Change*. Toronto, Oxford University Press.

Killewo, J., Anwar, I., Bashir, I., Yunus, M. and Chakraborty, J. (2006). "Perceived delay in healthcare-seeking for episodes of serious illness and its implications for safe motherhood interventions in rural Bangladesh." *Journal of Health, Population, and Nutrition* 24(4): 403–412.

MacDonald, R. (2005). "How women were affected by the tsunami: A perspective from Oxfam." *PLoS Medicine* 2(6): e178.

Mahmood, S. S., Iqbal, M., Hanifi, S. M. A., Wahedi, T. and Bhuiya, A. (2010). "Are 'village doctors' in Bangladesh a curse or a blessing?" *BMC International Health and Human Rights* 10(1): 18. www.biomedcentral.com/1472-698X/10/18.

Momsen, J. (2010). *Gender and Development*, New York, Routledge.

Nahar, N., Blomstedt, Y., Wu, B., Kandarina, I., Trisnantoro, L. and Kinsman, J. (2014). "Increasing the provision of mental health care for vulnerable, disaster-affected people in Bangladesh." *BMC Public Health* 14(1): 1–9. www. biomedcentral.com/1471-2458/708, accessed on 17/12/2014.

Narayan, D., Patel, R., Schafft, K., Rademacher, A., Koch-Schulte, S. (2000). *Voice of the poor: can anyone hear us?* New York, Oxford University Press.

Neumayer, E. and Plümper, T. (2007). "The gendered nature of natural disasters: The impact of catastrophic events on the gender gap in life expectancy, 1981–2002." *Annals of the Association of American Geographers* 97(3): 551–566.

Paolisso, M. and Leslie, J. (1995). "Meeting the changing health needs of women in developing countries." *Social Science & Medicine* 40(1): 55–65.

Parkhurst, J. O., Rahman, S. A. and Ssengooba, F. (2006). "Overcoming access barriers for facility-based delivery in low-income settings: Insights from Bangladesh and Uganda." *Journal of Health, Population, and Nutrition* 24(4): 438–445.

Parvin, G. A., Takahashi, F. and Shaw, R. (2008). "Coastal hazards and community-coping methods in Bangladesh." *Journal of Coastal Conservation* 12(4): 181–193.

Paul, B. K. (2010). "Human injuries caused by Bangladesh's Cyclone Sidr: An empirical study." *Natural Hazards* 54(2): 483–495.

Peters, D. H., Garg, A., Bloom, G., Walker, D. G., Brieger, W. R. and Rahman, M. H. (2008). "Poverty and access to health care in developing countries." *Annals of the New York Academy of Sciences* 1136(1): 161–171.

Phoenix, A. and Pattynama, P. (2006). "Intersectionality." *European Journal of Women's Studies* 13(3): 187–192.

Pitmman, P. (1999). "Gendered experiences of health care." *International Journal for Quality in Health Care* 11(5): 397–405.

Puentes-Markides, C. (1992). "Women and access to health care." *Social Science & Medicine* 35(4): 619–626.

Rahman, M. (2013). "Climate change, disaster and gender vulnerability: A study on two divisions of Bangladesh." *American Journal of Human Ecology* 2(2): 72–82.

Rahmatullah, M., Ferdausi, D., Mollick, M. A. H., Jahan, R., Chowdhury, M. H. and Haque, W. M. (2010). "A survey of medicinal plants used by Kavirajes of Chalna area, Khulna district, Bangladesh." *African Journal of Traditional, Complementary and Alternative Medicines* 7(2): 91–97.

Rashid, S. F. and Michaud, S. (2000). "Female adolescents and their sexuality: Notions of honour, shame, purity and pollution during the floods." *Disasters* 24(1): 54–70.

Rathore, F. A., Gosney, J. E., Reinhardt, J. D., Haig, A. J., Li, J. and DeLisa, J. A. (2012). "Medical rehabilitation after natural disasters: Why, when, and how?" *Archives of Physical Medicine and Rehabilitation* 93(10): 1875–1881.

Sen, A. (2001). "The many faces of gender inequality." *New republic* 18: 35–39.

Sen, A. (2013). "What's happening in Bangladesh?" *Lancet* 382(9909): 1966.

Shields, S. A. (2008). "Gender: An intersectionality perspective." *Sex Roles* 59(5): 301–311.

Stock, R. (1983). "Distance and the utilization of health facilities in rural Nigeria." *Social Science & Medicine* 17(9): 563–570.

Story, W. T., Burgard, S. A., Lori, J. R., Taleb, F., Ali, N. A. and Hoque, D. E. (2012). "Husbands' involvement in delivery care utilization in rural Bangladesh: A qualitative study." *BMC Pregnancy and Childbirth* 12(1): 28. www.biomedcentral.com/1471-2393/12/28.

Sultana, F. (2010). "Living in hazardous waterscapes: Gendered vulnerabilities and experiences of floods and disasters." *Environmental Hazards* 9(1): 43–53.

Uddin, J. and Mazur, R. E. (2015). "Socioeconomic factors differentiating healthcare utilization of cyclone survivors in rural Bangladesh: a case study of cyclone Sidr." *Health policy and planning* 30(6): 782–790.

Waldron, I. (1995). "Contributions of Changing Gender Differences in Behavior and Social Roles to Changing Gender Differences in Mortality." In Sabo, D. and Gordon, D.

F. (eds.) *Men's Health and Illness: Gender, Power and the Body*, Thousand Oaks, CA, SAGE Publications: 22–45.

Wiest, R. E., Mocellin, J. S. P and Motsisi, D. T. (1994). *The Needs of Women in Disasters and Emergencies*, Canada, Disaster Research Institute, University of Manitoba.

Zilversmit, L., Sappenfield, O., Zotti, M. and McGehee, M. A. (2014). "Preparedness planning for emergencies among postpartum women in Arkansas during 2009." *Women's Health Issues* 24(1): e83–e88.

7 Prevailing initiatives, gaps and people's expectations

7.1 Introduction: initiatives and gaps in disaster management

Bangladesh has been working on disaster management for a long time, about 50 years (Haque et al. 2012b), and is 'known for its continuous progress on disaster preparations at various levels' (Mutaza et al. 2013, p. 14). In particular, the country has evolved effective strategies to mitigate the impacts of natural disasters on human health (Cash et al. 2014), which can be observed from the substantially reduced death rate during cyclones (Fernandes and Zaman 2012 in Cash et al. 2014), which declined from approximately 500,000 deaths in 1970 to nearly 4234 in 2007 (Haque et al. 2012b; Islam et al. 2011). Robust policy, coordination at the highest levels, community engagement and localized decision-making have become vital approaches to policy in developing countries like Bangladesh (Haque et al. 2012). The Cyclone Preparedness Programme (CPP) was formed in 1972, and a Comprehensive Disaster Management Programme (CDMP) was established in 2004 to create additional institutional capacity and resources. In 2010, CDMP was expanded to cover five million people in high-risk communities and more than half the present five-year budget of CDMP was allocated to rural risk reduction and strengthening of community resilience (Cash et al. 2014). Again, disaster management is included in other national strategies, such as the poverty reduction strategy, to combat the direct effects of disasters on health and poverty (Government 2008). Thousands of cyclone shelters have been built, and Government, civil society and community stakeholders have mobilized strategies to provide early warnings and responses quickly after cyclone disasters (Cash et al. 2014).

All these plans and policies extended their threshold areas in the course of several years; for instance, most NGOs working in Barguna started after Cyclone Sidr. They are working on several disaster projects and training initiatives. Their initiatives to improve disaster preparedness and response were helpful during Cyclone Mahasen, increasing awareness of timely evacuation, but several problems revealed a gap between the prevailing plans and actual implementation at the local level. Alam and Collins mentioned in their research that 'despite good progress in cyclone preparedness, exemplified by the existing comprehensive disaster management policies of the Government of Bangladesh, localised vulnerability factors in cyclone hazards arguably remain only partly considered' (Alam and Collins 2010, p. 931). The present research identifies some questions and problems in implementing plans at the field level in light of responders' experiences of evacuation and emergency responses during Cyclone Sidr and Cyclone Mahasen. Here it should be mentioned that the data presented in this chapter come from interviews with government officials such as doctors, the Mayor of the Municipality, the Chairman of the Union and NGO officials who are work at the local level in Barguna. In addition, recommendations were made by both local inhabitants and these officials during the interviews and focus group discussions.

Preparations and taking safe shelter

Preparations for the cyclone were better planned and on a larger scale during Cyclone Mahasen compared to Cyclone Sidr. During Cyclone Sidr, warnings were broadcast and preparations undertaken immediately by the government at the national and local levels, with the help of NGOs. Volunteers worked at the field level, urging people to take shelter, and they helped older people and children reach safe shelters. In Tentulbaria village, Faruk and his wife, as trained NGO volunteers, tried hard to disseminate the warnings, take people to safe places and prepare the school building as a cyclone shelter. Faruk's efforts saved many villagers from Cyclone Sidr, though many inhabitants in Barguna municipality and even Tentulbaria village did not take shelter soon enough and had to face the devastating effects of the cyclone.

During Cyclone Mahasen, plans for warning dissemination, preparation and evacuation were better planned and improved (CDMP 2014). However, gaps have still been noticeable in implementing those plans at the field level. The number of evacuees among the respondents from the vulnerable locations and houses did not increase as expected. The following discussion explores the reasons behind this non-evacuation.

During Cyclone Mahasen, when the Bangladesh Meteorological Department (BMD) forecast the cautionary signals, Upazila Disaster Management Committees (UzDMCs) opened a control room on 14 May 2013 following the Standing Orders on Disaster (SOD) and being instructed by the District DMC (Disaster Management Committee). The UzDMCs instructed UDMCs (Union Disaster Management Committee) and CPP volunteers to disseminate the early warning. The Union Parsad (UP) chairpersons called an emergency meeting with UDMC members, CPP representatives, NGOs and local elites to make the necessary preparations to face Mahasen. Getting instructions from the UDMC chairmen, all UP members began to disseminate early warning messages in local communities using megaphones, microphones and the loudspeakers of local mosques. Though in most cases, people came to the cyclone shelter by their own arrangement, in some areas, for instance, in Badarkhali Union, UDMC arranged motor boats, rickshaw-vans and motorized vehicles to evacuate and help people to reach the shelters (CDMP 2014).

Besides the government warning dissemination, some NGOs in Barguna mentioned that they have their own contingency plan and separate cell and advisors for disasters. When any signal is forecast by the BMD, they open a control room and attend the emergency meetings called by the Government authority. They contact their trained caretakers of the cyclone shelters and volunteers by mobile phones.

Volunteers at the ground level are the most important part of warning disseminations and evacuations. They maintain very close coordination with UzDMCs and help local UDMCs to run the evacuation and response work successfully. There were 1,815 CPP volunteers in Barguna Sadar Upazila during Cyclone Mahasen, who were mostly local male and female inhabitants (CDMP 2014). I had an opportunity to talk with two male and female volunteers for the village in my study. They told me their stories of Cyclone Sidr and Cyclone Mahasen and the problems they faced while working as volunteers. Now the question arises, whether those preparations were enough to lead to a successful evacuation before the cyclone.

Were the cyclone warnings and awareness enough for successful evacuation before Cyclone Mahasen?

Compared to Cyclone Sidr, an increase in awareness and taking preparations can be seen among the inhabitants during Cyclone Mahasen. NGOs who worked on evacuation training could see the results of their work directly during Cyclone Mahasen. An NGO official mentioned in his interview, 'Our training showed its success during this Mahasen event. After receiving the signals, we could bring all the beneficiaries to the shelter with the help of the volunteers. Inhabitants could use the flags. The number of flags showed the intensity of the danger and males and females both could understand the message'. All of the training and experiences of Cyclone Sidr helped the respondents to become very aware of weather warnings and to draw up their own preparation plans for the next cyclone. They selected the place where they would take shelter during any further cyclone attack, and they tried to store some dry food. They became very alert. This improvement in preparation has also been mentioned in the CDMP report: 'The performance of all stakeholders at District, Upazilla and Union level related to early warning dissemination, evacuation, emergency preparedness and other necessary initiatives was found outstanding' (CDMP 2014, p. 1).

However, this improvement in awareness did not help to improve the number of evacuees during the warning periods. As discussed in the previous chapters, most of the urban respondents and some respondents in the rural study area did not leave their vulnerable houses during the warning period for Cyclone Mahasen. Several other factors created complex situations for the inhabitants, leaving them in confusion and facing dilemmas as to when and whether to take shelter before the cyclone attack. As Ahsan (2014) writes, 'an individual's risk preference is not only influenced by a single risky situation but also individual exogenous or background risks' and 'people always face multiple risks and such background risks may influence an individual's decision' (Ahsan 2014, p. 48). Compared to two respondents in Cyclone Sidr, only five respondents of the present study could leave their houses during the warning period of Cyclone Mahasen, though another

15 out of the other 25 respondents lived in tin-made vulnerable houses (Kutcha). They planned to take shelter at the neighbour's house rather than at the cyclone shelters. Many of them took shelter during the cyclone attack or had to stay at their vulnerable houses during Cyclone Mahasen. During another Cyclone Aila in 2009, a similar situation was also revealed. In his research, Mallick (2014) found that about 66% of the respondents (308 in number) were at their house or a neighbour's house during Cyclone Aila in 2009 in the Satkhira district, and only 14.1% took shelter at the cyclone shelter (Mallick 2014).

The main reasons behind the non-evacuation during Cyclone Mahasen include people's past experiences at the cyclone shelter (Mallick 2014), the lack of a cyclone shelter, an inaccessible location (Mallick et al. 2009; Paul 2009; Islam et al. 2011; CDMP 2014), the vulnerability of the shelter structure (Paul and Dutt 2010) and its dilapidated condition (Debnath 2007 cited in Haider and Ahmed 2014) and the low capacity of the shelter, which creates an overcrowded (Paul 2009; Alam and Collins 2010), filthy and unfriendly environment for all, especially for pregnant or sick women (Paul 2009), children, older and disabled people (CCC 2009).

Data from the interviews highlight the experiences of victims during the warning periods of Cyclone Mahasen.

Alauddin, a young man from the urban study area, described his worries about the nearest school building which is used as a cyclone shelter. After Cyclone Sidr, that school building has had one extra floor added to it without proper foundations. So, Alauddin and his family have taken risks by staying at their vulnerable house during the warning periods of Cyclone Mahasen, rather than gathering in that school building which might itself become vulnerable, being overloaded with evacuees. Alauddin explained, 'there is no proper cyclone shelter in the area . . . as one school building with one floor foundation has been converted into two stories, which becomes vulnerable being overloaded with people during disaster . . . we are afraid of taking shelter at that school and remain at home with high vulnerability . . . depending on fate . . .' (Male, age 27, urban area). Alauddin's reasons for being fearful show similarities to the situation mentioned in Cash et al.'s (2014) article: 'Although funds were quickly mobilised to help the most vulnerable people rebuild their home, an assessment noted that the new homes were similar to those that had stood before (rather than reinforced or strengthened), and would therefore be likely to suffer the same level of damage in future storms' (Government 2008 in Cash et al. 2014, p. 2098).

Again, confidence about their houses and preparations also discourages many respondents from evacuating during the warning time. People may feel their houses are 'quality houses' (Mallick 2014), strong enough for the cyclone attack, though most of the houses are made with tin and wood. Some of them were even reluctant to take shelter during the warning period of Cyclone Mahasen. Similar reluctance was also recorded among many residents of the Bolivar Peninsular in Texas, just three years after Hurricane Katrina, during the warning period of Hurricane Ike, which led to the deaths of over 100 people (Berg 2009; Kunreuther et al. 2009).

Nazem, who lives in Barguna municipality, stayed with his two young children and wife in his tin shade house during Cyclone Mahasen, as he felt 'my house is

built on a higher basement, it is strong . . . I rented it after Sidr for this reason . . .'
(Male, age 28, urban area). Even Rima, Karim and more from Tentulbaria village
stayed at home during Cyclone Mahasen. But staying in vulnerable houses did
not turn out to be a good decision for most of the respondents. Many of them had
to leave their house during the cyclone attack in strong wind and storm waters or
else had to face injury and panic while staying inside their damaged house.

Therefore, many of them took the same risk again in Mahasen, as in Sidr. It
could have led to a large number of injuries during Cyclone Mahasen, if it had
turned into a severe cyclone. The question arises: if all respondents from vulnerable locations wanted to take safe shelter during Cyclone Mahasen, was it
possible to provide them with enough shelters? There were only 70 shelters in
usable condition in Barguna Sadar Upazila with capacity for 30,000 people during
Cyclone Mahasen (CDMP 2014). The total population of Barguna Sadar Upazila
is 237,618, and the rural population is 210,659 (BBS 2006). So, the capacity of
the shelters is far below the amount needed. For instance, Tentulbaria village has
only one cyclone shelter with a maximum capacity of 1,000 people, but the total
population of the village is 3,345 (BBS 2006). This village does not have any
multi-storey buildings, except one or two private houses. So, an insufficient number and unsuitability of the cyclone shelters make the success of the preparation
and evacuation plans questionable.

Medical responses: preparations and gaps

The government has specific plans for the medical relief (Government 2008).
During Cyclone Sidr, the government was prompt to deploy medical personnel
to the affected districts. Initially, medical personnel from public sectors in neighbouring districts were sent. Additionally, the military set up their own medical
camps, and many civil-society groups, NGOs and other organizations sent their
medical teams. For instance, BRAC (Bangladesh Rural Advancement Committee) responded with 240 local community health workers just hours after the
storm passed (Ahmed 2007 in Cash et al. 2014). The International Federation
of Red Cross and Red Crescent Societies assisted more than 83,000 people with
basic healthcare and emergency medicines in 13 districts, through the Bangladesh
Red Crescent Society's (BDRCS) mobile medical teams during the emergency
phase (IFRC 2010). The government had obtained medicines and accessories
worth US$6.8 million within a few days of Cyclone Sidr, accessing funds from
the health sector-wide programme (Health, Nutrition and Population: HNPSP)
(Government 2008).

During Cyclone Mahasen, the Government prepared 1,300 medical teams for
all the affected 47 upazilas. 'Medical teams were formed for each union, con-
sisting of one medical officer/representative from Upazila Health Complex and
members of the union community health clinics' (CDMP 2014, p. 2). Fire brigade
teams were kept ready for the emergency, and regular contact was maintained
with the centres at the field level. In addition, the Bangladesh Navy, Air Force and
Army were prepared for emergency operations and responses. Now, if we focus
on the ground level and consider the opinions of the respondents, it is revealed

that the medical response was not enough for the injured victims. None of them were checked by a doctor or medical staff at their locality; rather, they had to visit distant healthcare centres. The following section discusses the gaps revealed in the field.

Were healthcare preparations enough for all the victims in Barguna?

In Barguna, according to government officials, medical teams and items were ready three to four days before the cyclone attack as part of healthcare preparations for the Cyclone Sidr emergency response. Contact was conducted with all the health centres at the field level until the mobile network stopped working. During the post-disaster period, 46 medical teams worked beyond the usual arrangements in Barguna. They had enough medicines and healthcare infrastructure to provide services to the victims after the cyclone. According to the government officials, healthcare preparations were satisfactory in Barguna, 'We were prepared, we had sufficient infrastructure . . . we did not face lack of beds of the hospitals or infrastructure. We opened the closed community clinics after cleaning them and made them prepared' (Government Official). But this quotation does not represent the full picture of what happened in Barguna; it rather shows one side of the preparations and gives rise to questions about the health provisions during the post-disaster period. First of all, victims of the remote rural Barguna did not get any medical relief for two to three days after Cyclone Sidr, whereas, to reduce the consequences of disasters, drinking water, food, healthcare and other necessary relief services should be distributed to the most affected people (Shultz et al. 2005). Interviews with government officials revealed that doctors and medical services from the Barguna Municipality did not reach the remote and highly affected areas until two to three days after Cyclone Sidr. Even then, in some areas, doctors in charge of rural medical centres also could not reach the affected areas in time. Some distant healthcare centres were without any doctors to deal with the emergency.

Secondly, many areas of rural Barguna did not have active healthcare centres at the nearest location, and the transportation system (which was difficult prior to the cyclone) became even worse after the disaster. So, victims from the remote areas could not reach the distant healthcare centres until transport became accessible. According to the hospital officials, 'Injury cases increased slowly at the hospital after Cyclone Sidr'. Therefore, the preparations of the medical sectors were most helpful to the victims of the remote areas like Tentulbaria village only if and when they could reach those distant healthcare centres. For instance, Kamal and his brother Faruk (respondents of the present research) received delayed treatment for their injuries once they reached distant healthcare centres. By this time, the delay in treatment had made their health conditions worse.

Thirdly, injured urban victims mentioned in their interviews that they found the urban healthcare centres overflowing with patients – a situation which was denied by Government officials. Again, all injuries could not be treated in most centres of Barguna and so some were referred to Barisal and Dhaka. Preparations at the urban healthcare centres were better than usual, but not sufficient for the higher needs during disasters. Moreover, these preparations were not helpful to those victims living in the rural areas such as Tentulbaria village.

What happens two to three days after Cyclone Sidr?

NGOs and INGOs reached Barguna with their medical teams at an increasing rate after two to three days of Cyclone Sidr (Source: Interviews). NGOs who owned healthcare centres in Bangladesh arranged the medical teams. They worked in Barguna for different periods of time, from seven days to several months. They also worked according to their capacity in specific areas. According to an interviewed NGO official, 'We decided to work in two unions because of the location and transport system. We had our own limitations also'. This NGO could arrange a medical team, including three doctors, after two to three days. They offered primary treatment and had medicine for diarrhoea, dysentery etc. But these facilities were not available all over Barguna. Many areas remained uncovered by the medical teams after Cyclone Sidr. In the study areas of the present research, no respondent was checked by any visiting medical staff, not even in the worst-affected Tentulbaria village. As discussed above, distant rural healthcare centres in Barguna had limited services for certain types of injury, and, again, absence of staff intensified the problem.

However, there was success in preventing epidemics after Cyclone Sidr and Cyclone Mahasen. According to a Government Report, 'relief efforts have concentrated on measures to prevent and control outbreaks of disease. A basic disease surveillance system has been created in and around cyclone-affected areas' (Government 2008, p. 39). Additionally, the Government of Bangladesh, with several multilateral partners and civil-society groups, focused on the availability of safe water and proper sanitation systems (Cash et al. 2014). Water purification tablets and oral saline were distributed with relief items. NGOs helped to treat contaminated ponds in the most severely affected areas to assure the availability of clean water and prevent waterborne diseases (Ahmed 2007). In Barguna, NGOs (Save the Children) installed water purification machines (four to five in number) in the affected areas, and public tube wells were washed. People could collect safe drinking water. All these activities prevented the occurrence of epidemics (Source: Interview with Govt. official).

Another important aspect of disaster consequences, the mental health needs of the survivors, did not receive sufficient attentions by response programs during the post-cyclone periods. As Nahar et al. (2014) mentioned in their research, 'the serious and widespread mental health consequences of natural disasters in Bangladesh have not received the attention that they deserve' (Nahar et al. 2014, p. 1). There was no notable assessment conducted by the government. According to a government report published after the five months of Cyclone Sidr, 'Many people are still missing and thousands of others have sustained physical trauma and mental setbacks; it is impossible to quantify these associated losses at this time' (Government 2008, p. 39). Like most of the affected coastal areas, there was no major response programme on mental health after the cyclones. Only some NGOs have taken the initiative to provide psychological support a few months after cyclone

Sidr. BDRCS sent a psychological support delegate to the Sidr operation centre in Barisal in March 2008, and they made field visits to some affected areas, including Barguna (IFRC 2010), but not in the study areas of the present research. At the local level, an NGO worked with traumatized child victims of Cyclone Sidr in a three-month programme. Though their programme resulted in improvement among the children who participated, they did not continue the programmes for long.

Challenges faced by relief distribution

The Government of Bangladesh, in collaboration with the United Nations, the National Red Crescent Society and the IFRC, undertook several humanitarian assessment missions and relief operations after Cyclone Sidr. The government allocated BDT 450 million (US$6.7 million) for relief distribution, sending medical teams and housing construction. The government and international and national organizations distributed food items among the 2,190,017 families affected (Government 2008). For instance, the IFRC distributed relief (food and nonfood) among more than 84,000 affected families (IFRC 2010). Additionally, the international community rapidly responded and pledged approximately US$241.7 million in aid (Government 2008). At the field level, emergency responses were also implemented in Barguna for several months by the government and NGOs. One of the interviewed local NGOs mentioned, 'We, with our own limited resource and man-power, we responded to Cyclone Sidr (covering 72381+42643 families; supported by WFP, UNICEF)'.

During Cyclone Mahasen, UzDMCs stored the food from the local supply depots after receiving the warnings. Barguna Sadar UzDMCs had distributed 72 MT rice and BDT 400,000 as relief among the victims (CDMP 2014), though it has been reported that this last-minute preparation created chaos among the stakeholders. Local NGOs mentioned that during Cyclone Mahasen, the food items they had previously ordered from the local market had been taken by the government's authority. They had to rebook other items for their beneficiaries.

Several international and national NGOs also provided significant assistance to the relief distribution. For instance, the UNDP provided assistance to 2,200 affected families in Bhola and Barguna District. BDRCS and the IFRC assisted 4,000 households in three districts, including Barguna. The World Health Organization, through the Directorate General of Health Services, provided financial assistance for disease surveillance and medical kits amounting to BDT 4 million to support affected victims. UNICEF distributed 5,500 family kits, including hygiene kits (DDM 2013).

Here it should be mentioned that all NGOs interviewed in Barguna had emergency response and recovery plans to distribute relief, but providing healthcare facilities for the injured victims was not included among those plans. Mainly the government, but very few NGOs, in Barguna could arrange medical teams for injured victims. The following examples of two large NGOs (according to a number of staff and partners) in Barguna are helpful to understand preparations for Cyclone Mahasen.

NGO 1

This NGO had a prepared plan before Cyclone Mahasen. About 50 trained staff (including staff from the partner NGOs) were ready to work during disasters. Approximately 60% of the people in the field as the volunteers were women. Non-food items were ready for 2,500 families. The morning after Cyclone Mahasen, a rapid needs assessment was conducted. Three days after the cyclone, relief was distributed among 2,050 families. The study area of this research, Badarkhali union, received relief for 500 families. They also had recovery plans including cash for work and training.

NGO 2

After the cyclone warning was forecasted by the government, this NGO opened a control room and disseminated the warnings to the field levels. They kept meg-aphones, microphones, motorcycles and cash in hand: Tk 500,000 ready. They stored foods (dried and mixed) and medicine and kept ordering from the local markets. They booked trawlers, boats and speedboats (five to seven in number). They planned where they would keep their food stored at the field levels. They had a few prepared cyclone shelters where both people and cattle could take shel-ter separately. Some shelters have toilets, tube wells and water tanks, and cyclone shelter managers had first aid boxes with them.

NGOs also prepared relief items, with some common items being dry foods, clothes, some daily accessories, sanitary napkins, oral saline and water purifi-cation tablets. A list of relief items which were prepared by one NGO during Cyclone Mahasen have been included in Table 7.1, 'Relief items per pack in Cyclone Mahasen'.

The question arises: could the responders distribute the relief supplies satis-factorily, or did they face challenges? These questions were asked during the interviews of the government and NGO officials who worked in Barguna dur-ing Cyclone Sidr and Cyclone Mahasen. They mentioned several factors and constraints which made their relief work challenging during the cyclones. The following section will focus on these challenges and highlight the gaps in the prevailing disaster management plans.

Emergency response after disasters becomes a great challenge for the respond-ers for several reasons, including an inherently chaotic post-disaster environment, lack of sufficient resources, coordination difficulties among the actors and lack of proper disaster management plans. Although it is assumed that developing poor countries often require efficient international ERCs (Emergency Relief Chain), this point is not limited to them, as seen, for example, in Hurricane Katrina and the 2009 bushfires in Australia. These disasters established that even pre-planned and well-developed evacuation and response plans can fail during major disasters and that wealthier countries can improve their disaster response and provisions for emergency relief (Oloruntoba 2010). Comparing several cases, it is revealed that different constraints on relief work are experienced during disasters all over the world. For instance, Litman 2006 mentioned in his research that Hurricane

Table 7.1 Relief items per pack in Cyclone Mahasen

SL.NO	Particulars	Specification	Number of items
1	Toothbrush	Standard (2 adult + 3 children)	5
2	Toothpaste	120gm	1
3	Bathing soap	150gm	2
4	Washing soap	130gm	1
5	Clothes washing powder	500gm	1
6	Bath towel (gamcha)	53 X 23 inches	2
7	Comb	Standard medium size	2
8	Coconut oil	400ml bottle	1
9	Candle (6 pieces)	Large size (200–250gm)	1
10	Lighter	Sunlight	1
11	Sanitary napkins	10 pieces, standard	2
12	Towel	Standard (Gemini)	1
13	Sari (Women's cloth)	Standard size (Pride)	1
14	Lungi (Men's cloth)	Standard size (Pride)	1
15	Bed sheet with pillow cover	Standard size (Classic Home textile)	1
16	Blanket	5' X 7' size. 800–900gm	1
17	Nail cutter	Standard size	1
18	Plastic mug	1.5 litre capacity	1
19	Plastic bucket with lid	20 litre capacity	1
20	Paper cartoon with plan logo	Mentioning items name printed	1

Katrina's evacuation plan functioned relatively well for motorists but failed to serve people who had to depend on public transit, whereas Hurricane Rita's evacuation plan failed because of excessive reliance on automobiles, resulting in traffic congestion and fuel shortages. Again, during the response period of Hurricane Katrina, it was not a lack of resources but lack of better planning which left tens of thousands of people without food, water, medical services or public services (Murdock 2005; Litman 2006). 'Civil organizations were not allowed into the city to provide assistance' (Litman 2006, pp. 11–18), and all human resources and relief materials were ready for deployment, but they were 'turned back, misdirected or misused' (Murdock 2005 cited in Litman 2006, pp. 11–18).

The constraints on a successful response programme were different in Taipei after Typhoon Nari. During this typhoon, the modern healthcare system of Taipei was seriously damaged, which prevented the provision of medical services at full capacity, and even some emergency services were closed to the public. Hospitals slowly restored medical services to the community, which in some cases took one and half months (Lai et al. 2003).

Again, during the Bam earthquake of Iran in 2003, 'the main obstacles to pre-hospital medical response were the lack of a disaster management plan, the absence of disaster medical assistance teams and the overall lack of resources' (Djalali et al. 2011, p. 9). Transporting injured victims was also a great challenge for the response programmes.

But obstacles to the response programme were not same in Myanmar after Cyclone Nargis in 2008. Cyclone Nargis killed tens of thousands of people and left hundreds of thousands of people homeless and vulnerable to injuries and diseases (Stover and Vinck 2008). Successful response after this mighty cyclone was not possible for the country alone, and yet Myanmar's commander-in-chief declared that Burma was capable of handling the relief effort. But the government failed to take essential measures during the critical days and weeks after the cyclone. Again, the Government allowed limited international assistance, for instance, the Myanmar Red Cross Society worked in a few affected townships, whereas 'scores of international aid workers remained grounded in neighbouring Thailand waiting for visas' (Stover and Vinck 2008, p. 730).

Now, if we look for a successful story of a response programme, relief efforts after Cyclone Larry of Australia provide one case, as 'one of the more effective in the history of emergency cyclone response in northern Australia' (Oloruntoba 2010, p. 85). The key success factors in Cyclone Larry were 'the preparedness and readiness of the authorities and the affected communities' and the 'unity of direction and cohesive control of responding government agencies' and their 'alliance with the private sectors' (Oloruntoba 2010, p. 94).

The above discussion shows that success or failure of response programmes after disasters depends on several factors, such as the socio-economic and political conditions of the country, prevailing disaster management plans and, of course, the severity of the disaster. Even where preparation has taken place over decades, and disaster management plans made, they can face several challenges at the field level while being implemented, for instance, in Bangladesh. Though Bangladesh has improved its disaster management plans, response programmes still face several major constraints during implementation. Government and NGO officials in Barguna faced several challenges, obstacles and problems regarding relief distribution during Cyclone Sidr and Cyclone Mahasen. Their experiences reveal the significant gap between the prevailing plans and their implementation.

DISRUPTED TRANSPORT SYSTEM

Disrupted transport systems become a major constraint on the implementation of rescue and response plans and programmes after disasters (Ahmed 2007). In the present research, all respondents, inhabitants, Government and NGO officials repeatedly mentioned this constraint. According to the responders, they were prepared with vehicles, relief and medical teams, but they could not reach the remote areas until the roads were cleared of fallen trees. Clearing roads took two to four days, which delayed the response programmes. For these days, many remote areas became totally disconnected. Waterways were widely used historically in Barguna to reach remote areas, but after the construction of roads, many waterways were blocked, and roads become the only ways to reach those areas. Again, some remote areas, like Badarkhali union, have a very poor transport system. There is no paved road but only earthen roads interconnecting the villages. These earthen roads become very difficult to use after the cyclones.

So, medical teams or respective doctors could not reach the farthest areas or even be present at their own work places due to damaged roads. According to the Government officials, they waited two to four days until they could provide treatment for injuries in the remote areas of Barguna. Here a question arises: if medical teams, with all their preparations, could not reach the remote areas of Barguna, then how could a poor, injured or sick inhabitant without any help reach the furthest healthcare centres after the cyclones? This situation created an impact on the admitted injury cases in the healthcare centres in Barguna, which increased slowly as the days passed. These delays worsened the injuries or sickness (see discussion of this in Chapters 3 and 4). So, medical teams might be ready in Barguna Municipality, but if these services could not reach the affected areas during the emergency periods, then a question arises about the success of the response programmes.

Again, as well as medical help, relief with food, water purification tablets and oral saline could not be distributed in time after Cyclone Sidr due to the damaged roads. Tentulbaria village did not get any relief for four to seven days. One of the NGO officers mentioned that 'we reached those remote areas with cooked foods [Kichuri, food like porridge] after three days of Cyclone Sidr. We saw aged inhabitants starving for days. They were so hungry that they collected leaves as food . . . They did not have anything else left as food . . . the experiences were unforgettable'.

Even six years after Cyclone Sidr, in some areas of Barguna, conditions had not improved during Cyclone Mahasen. The furthest remote areas received delayed relief and even had no help at all for days. The experiences of the NGOs indicate conditions after Cyclone Mahasen:

> We, my colleagues and I, tried to reach the affected areas just after Mahasen but could not go further after 2 km; the roads were all blocked by fallen trees. We had to return. Those remote areas could not be reached before two days. There were even some remote areas where no connection could be made for ten days. Those areas mostly did not have enough food reserves and desperately needed help. If anything could not be reached, those inhabitants will be without food for all those days.

NGOs need to plan to store foods in remote areas. But this arrangement is not possible for most of the NGOs in Barguna. An NGO staff member mentioned that 'we have to choose those areas where we can take the truck with relief goods'. So, the villages of Badarkhali union did not get any help from this NGO, as they do not have a road network.

LACK OF LOCAL FACILITIES

While the number of beds and medical staff are quite inadequate to face the high demand after cyclones, the provision of injury treatments in the Barguna healthcare centres was also inadequate. Most of the healthcare centres in Barguna could not provide treatment to critical patients, and the conditions of rural healthcare centres were even more inadequate in facing the increased demand after the disasters. These healthcare centres have several problems such as 'bed occupancy far higher than safe thresholds, chronic shortages of doctors and nurses and infrastructure deficits related to electricity, transport and water' (Huque et al. 2012 in Adams et al. 2013, p. 2105).

In some areas of Barguna, NGOs helped to transfer injured patients to the hospital (Ahmed 2007), but transferring a patient is very difficult from the village, especially after disasters. An NGO officer who works in Badarkhali union mentioned that in his experience, '[t]hose villages do not have a sufficient number of community clinics. Patients from Tentulbaria and Gulshakali have to visit Phuljhuri which worsens the health condition of the patient. Patients do not remain under the primary treatment while visiting Phuljhuri from Tentulbaria'.

On the other hand, electricity and water supplies become a great problem for healthcare centres to continue their activities after disasters. After Cyclone Sidr, the electricity supply returned to normal in Barguna only after one month in some locations – and in some areas, it took even longer (Government 2008). After Cyclone Mahasen, it took one week. According to a member of the medical staff, 'a lack of electricity and water became a great problem after Cyclone Sidr rather than the lack of medicines'. It became very difficult to preserve important medicines and continue treatments without electricity. The lack of electricity also interrupted mobile connections, as mobiles could not be charged properly, which also disrupted the response programmes (Ahmed 2007 cited in Cash et al. 2014).

LIMITATIONS AT THE LOCAL LEVEL OF DISASTER MANAGEMENT

Though the Bangladesh government has an excellent disaster management plan for cyclones, there are some gaps and limitations revealed at the local level which became obstacles for successful implementation. First of all, local government as a UDMC does not have any budget allocation for emergency purposes (Government 2008; CDMP 2014). Mostly, they do not have any equipment, vehicles or boats for evacuation or early warning dissemination. All necessary arrangements for the preparation during Mahasen were 'made by the personal donation of UP chairman and members' (CDMP 2014, p. 4).

Furthermore, interviews with local government staff revealed that some severely affected unions/villages do not get proper attention from the national government. This problem was also mentioned in the Bangladesh Report (2013): 'It has been observed that some severely affected villages are not included in the response and recovery efforts as they do not fall within the most affected unions or upazila. This has resulted in most deserving families to be left out of the response and recovery programmes' (DDM 2013, p. 49). Again, this exclusion brings great problems for the elected authorities, who have to face the social pressure of helping people during the cyclones. Lack of any budget creates difficult challenges for those local authorities to implement plans. This weakness and inadequacy of local government also create problems for the NGO projects, who emphasized the development of strong linkages among communities and government departments (Murtaza et al. 2013).

Secondly, interviews revealed that evacuees take shelter during the warning period, mainly one day before the cyclone attack, but in many shelters, there is no arrangement or plan to distribute foods to those evacuees (CDMP 2014). It then becomes a pressure for the NGOs who helped the people to take shelter. Evacuees including children, older people, pregnant and lactating women need food at least three times before the cyclone attacks, but this situation is difficult to explain to the authorities, the Government and even donors. In most cases, the authorities become involved only after the cyclone hits the areas.

Thirdly, the local volunteers who help evacuate and rescue the victims do not have the proper tools. It has been mentioned in other research that volunteers had been provided with raincoats, gumboots, hardhats, life jackets and flashlights to facilitate their activities in adverse weather (CPP and BDRCS 2007 cited in Paul 2009), but volunteers in the present study areas were not well equipped with the necessary items. They had received some materials a long time back, and if they lost those in previous cyclones, they were not replaced. They worked without any volunteers' gear. These problems were also identified in the CDMP report (2014) on Cyclone Mahasen. It has been described that 'Some of them [volunteers] are wearing only an old vest with a logo of Bangladesh Red Crescent Society' (CDMP 2014, p. 5) – this reflects the experience of Faruk, a respondent in the present research. Faruk and other volunteers of the village did not have any microphone, megaphone, sirens or radio. They had to use their personal mobile phones to maintain occasional communication with union team leaders. But having no support from anywhere, it was not always easy for the volunteers to fulfil their responsibilities during the cyclone warning periods. Poor volunteers did not get any compensation for their injuries. Faruk got injured while disseminating Cyclone Sidr warnings but did not get any help to complete injury treatment. He still suffers from his injury pains.

ADDITIONAL CONSTRAINTS FOR NGOS

NGOs working in disaster management face several additional problems. During the interviews, NGO officials mentioned their limitations in providing help for all victims due to lack of budget and manpower. Most NGOs rely on donor funding and 'cannot initiate a disaster response before funding becomes available' (Seaman 1999 cited in Balcik et al. 2010, p. 23). Most of the time, they cannot extend their budgets. Several quotations illustrate their situations, 'We have to select the poorest of the poor to distribute the relief aids' and 'It would be better if we could help all the victims'.

Due to a lack of budget, NGOs sometimes cannot even deliver essential tools, such as first aid boxes or volunteers' gear, after awareness training. As NGO officials mentioned in the interviews, 'If we cannot supply them essential kits, how will this training be successful?' This opinion was also voiced by the respondents of Tentulbaria village, who mentioned that they received training from NGOs but not any essential kits. They became vulnerable when rescuing victims during disasters, even taking risks themselves to help evacuees to reach safe shelters.

Besides the budget limitations, NGOs mentioned several other constraints, including insufficient help from the local government, mismanagement in government plans and implementation, and political and illegal local pressure to select the victims.

COORDINATION DIFFICULTIES, NGOS AND RELIEF DISTRIBUTION

'For disaster management, several inter-related systems need to work together in time of crisis, with very little advance notice' (Cash et al. 2014, p. 2097). Many factors contribute to coordination difficulties in disaster relief, such as the inherently chaotic post-disaster relief environment, the large number and variety of responders in relief distribution, and a lack of resources (Balcik et al. 2010). Besides this, lack of coordination in material and information flows among the responders also becomes an important factor for implementing response plans. Unavailability of information has been identified as one of the constraints by the NGO officials interviewed in Barguna. They mentioned that they could not find respective Government staff to provide essential information on the local residences which was needed during the relief distribution after Cyclone Mahasen. Though the Directorate General of Health Services (DGHS), Ministry of Health and Family Welfare (MoHFW), ordered all health and family planning service providers to stay at their respective offices, and all leave of health service providers was postponed during the cyclone warning periods, they were not found at their offices. Again, members of UDMCs were not well aware of their Standing Orders for Disasters (SOD) appointed roles and responsibilities (CDMP 2014; DDM 2013). Besides this, negligence of duties was also found among some of the members. Government workers and staff mostly live in the urban areas, and they visited the affected rural areas only after several days. One NGO officer stated, 'Government staff should have more concern but they were absent from the field . . . taking a rest at home due to weekly holiday . . . but does emergency follow any holiday? Government staffs mostly do not have any liability at the field level. One NGO cannot do everything. Government has enough infrastructures but they (the government) remain slow in regards of disaster responses'. These coordination difficulties have negative consequences in response programmes. As Simatupang puts it, ' lack of coordination among chain members has been shown to increase inventory costs, lengthen delivery times and compromise customer service' (Simatupang et al. 2002 in Balcik et al. 2010, p. 22).

Again, 'data for risks associated with morbidity and mortality from natural disasters should be obtained as soon after the event as possible to inform planners what needs to be done to mitigate the effect of the next natural disaster' (Cash et al. 2014, p. 2094). But delays in damage reports four months after Cyclone Mahasen were reported by the interviewed NGOs, and the Department of Disaster Management mentioned this delay in their report (DDM 2013). NGOs in Barguna reported this to the high levels of government, but there was no improvement made. As a result, in some cases, donors did not agree to help during Mahasen. As an NGO official says, 'How much can an NGO do if government does not focus on the implementation of the plans. NGOs should help the government, but the government lags behind in the field'.

Again, there were differences between field information and Government information, which became constraints for the NGOs in relief work. This difference is mainly created by delays in field surveys conducted by government officials after the disaster. One of the government officials mentioned in his interview that 'we could not reach the distant remote areas even 3–4 days after Cyclone Sidr'. In addition, negligence of duties, irregular presence in the field office and corruption amongst government officials create anomalies in data and data collection after disasters. One respondent said, 'there were three government officers in my native village; they left the area during the cyclone attack and did not return soon' (Male, age 45, urban area). Another respondent from the urban area said, 'our chairman, commissioners all are corrupt . . . they only think of their relatives and supporters of their political parties . . . they provide the wrong information about victims' (Male, age 52, urban area). For instance, the Shelter Coordination Group (SCG) conducted an assessment in Barguna after Cyclone Sidr in February 2008 to look at the needs of the affected people in regards to Early Recovery Programming. The report mentioned that one of the main challenges was the lack of clear and authoritative data on the extent and characteristics of damage to housing from Cyclone Sidr. Additionally, the result of the assessment shows 'slightly more damage to housing than the government has reported' (SCG 2008, p. 8). Besides these differences, examining the reason behind disaster morbidity is problematic in Bangladesh because data collection is often episodic and has become more routine in the past few years. Many people are unidentified in Bangladesh's national birth and death registries (Cash et al. 2014), and 'baseline data on pre-disaster status are also rarely available, which acted as a barrier in determining the actual impact of the tropical storm Mahasen' (DDM 2013, p. 49).

Again, lack of cooperation was also revealed during the emergency periods. There is illegal pressure to distribute assistance selectively, among more well-known and well-off families, who do not fulfil the criteria for NGO beneficiaries. But it was very difficult for most of the NGOs to face these pressures and stick to their plans. Again, they could not do any pilot surveys because it is difficult to do pilots or tests during emergency periods (Cash et al. 2014). The following quotations from interviews illustrate this.

> NGO 1: Making the family list (register of those who are most in need of aid) is a great challenge as the Member and chairman got involved and the process was corrupted. Persons who should not get on to the receivers list were included. We tried to recheck with our own staff but still they got on to it. During emergency it was also very hard to justify everything. This was the biggest challenge.
>
> NGO 2: They pressure NGOs to give relief to the chairman, lawyers, government officers first. They say, 'The poor did not lose anything, they did not have anything. So, they do not need anything'. We replied to them, you can recover but they (the poor) will not. We had to struggle to distribute relief, we had to face constraints to distribute legally. When conducting emergency operations we have to be strong, facing political pressure, being threatened with looting. We have to be honest to fulfil the criteria to distribute properly. We have to take our own risks. I have been arrested three times as I did not listen to these powerful local people. But they could not file a law-suit against our NGO.

Beside this pressure, permission is needed from the government to distribute the relief, but this process created delays because of differences among the government officials.

POLITICAL PRESSURE, NGOS AND RELIEF DISTRIBUTION

Political pressure was a great problem for proper relief distribution in Barguna during previous cyclones. Sometimes, political pressure from a higher level makes the local authority helpless. Hossain et al. (2008) commented that '[the] Chairman may be a figure of great authority with the local village thief, but he is powerless against criminals from further afield' (Hossain et al. 2008, p. 2). For instance, political pressure became so problematic for one NGO that they had to cancel relief distribution after Cyclone Mahasen: 'We had to postpone distribution in one union. Two political parties started conflicting over relief distribution and we were threatened that they would not let us enter the area. We proposed to them to decide themselves but they could not take decisions even within seven days. So, we had to postpone aid to that area and distribute to the other areas to save our food items'. However, the question then arises about poor victims who need immediate relief to survive after disasters. The ultimate result of political conflicts is that they suffer and are made more vulnerable. Here, it is worth reflecting that '[v]ulnerability can be traced back to quite remote roots and general causes that entail socio-economic processes and political factors, which are requisite for understanding why hazards affect people in varying ways and why people experience disaster differently' (Ray-Bennett et al. 2010, p. 581).

7.2 Gender, culture and the gap between planning and implementation

The Bangladesh government formulated the 'National Plan for disaster Management 2008–2015' in 2008 and 'Disaster Management Act, 2012', both of which have gender-related issues as an objective for overall development in disaster management (Ministry of Finance 2014, p. 126). Government plans give priority to women in evacuation, taking shelter and relief distribution in disasters (MoDMR 2012). However, as this present research reveals, there are a number of different factors, constraints and challenges which are influencing the actual implementation of these plans.

Evacuation plans and cyclone shelters

To start with the evacuation plans, according to the government report, 'Women, children and handicapped people are taken to the safe shelters on a priority basis at the beginning of any serious disaster and they are given food and medicines' (Ministry of Finance 2014, p. 126). However, this did not occur in Barguna during previous cyclones. The disaster managers have given different opinions about this. Most of them, especially local government officials, mentioned that their preparation was not sufficient for the female evacuees during the previous cyclones, due to a lack of human and financial resources. Women and children should have received more attention, especially in regard of food and shelter, but there was no vehicle to carry them to the safe place. Besides, river islands in Barguna do not have proper healthcare facilities for pregnant mothers, and it was not possible to arrange any special healthcare facilities for them at the cyclone shelters, even during Cyclone Mahasen in 2013.

According to the 'Cyclone Shelter Construction, Maintenance and Management Guideline 2011' the location of shelters should be within 1.5 km, in close proximity to the vulnerable community, so that evacuees may easily reach the shelter during the warning period and road communication to shelter must be suitable for all the communities including women, children, aged and the disabled (MoDMR 2012, p. 11). But the implementation of this plan was not always observable in the field. Sara, a respondent in Barguna, said, 'we have a cyclone shelter within 2 km but we have to cross a river . . . it is not safe to cross the river during the warning period and if we take the road, it will take a very long time, crossing a long distance' (Female, age 24, urban area). So, Sara and her mother did not evacuate during the Cyclone Mahasen warning.

Lack of separate space and arrangements for women at shelters became major reasons for the avoidance of cyclone shelters by the female inhabitants of Barguna. NGOs working on the evacuation mentioned that vulnerable people (older people, pregnant women and new mothers) did not get any special care at the cyclone shelters. Though government guidelines mentioned that shelters should have separate rooms and toilet facilities for women, especially for pregnant women, during Sidr no female respondent could access such facilities. Six years after Cyclone Sidr, some cyclone shelters did arrange separate space for women but pregnant women and new mothers did not get any priority. NGO officials suggested that this issue should be discussed and dealt with during disaster planning, because during Mahasen, 70–75% people came to the cyclone shelter, and among them, 40–45% were in these special groups (older people, pregnant women and new mothers), but they did not receive any special treatment. Pregnant women and new mothers risked their lives during Cyclone Mahasen because of this lack of facilities in the cyclone shelters.

Muna, a young mother from the urban area, described her bad experiences during Cyclone Sidr, which then compelled her to stay at her very vulnerable house with her three-month-old baby during Cyclone Mahasen. She described the environment of the shelter,

> people of different socio-economic levels, high to lower level Government officers and ordinary people [slum dwellers and general] took shelter in the nearest Government office. Government officers, who work in that office, got good separate rooms, whereas general people of all levels took shelter in one place in overcrowded rooms with their hens and ducks. The environment was not good and there was no food and healthcare facility in the shelter for the patients. There was a toilet facility but it was overcrowded. The windows of the building were broken because of the strong wind. . .
>
> (Female, age 25, urban area)

Having had these experiences in Cyclone Sidr, and as conditions did not change, Muna did not take shelter in that same facility during Mahasen with her young child. She waited at her house to take shelter in a neighbour's house. But being scared by the increasing wind and rain, Muna took shelter twice in one night at her neighbour's house. She and the young baby were made vulnerable by moving from the house during strong winds. But Muna was not comfortable to stay the whole night in her neighbour's house, and so, in her dilemma and panic, she left her house twice that night. Muna's experience of taking shelter is similar to those of Fulon and other respondents of the present research. This hesitation was also mentioned in Custer's (1992) research on 1991 Cyclone in Bangladesh (to see more, Custer 1992), but unfortunately, those social constraints of disaster preparation have not changed even after 25 years.

Healthcare facilities for women and cultural barriers

The Bangladesh Government's plan is to provide necessary medicines to the most vulnerable groups, including women at the cyclone shelter, but this was not possible in the remote areas of Barguna. Local Government officials and NGOs in Barguna have mentioned that preparations were not good in the rural areas, though in some urban centres, preparations were satisfactory. According to the government officers, 'prevailing facilities were enough for women, I did not feel the need to have special arrangements for women during a disaster'. In one way, this is true, as Barguna municipality arranged healthcare facilities beyond their capacity, but the problem arises for the rural inhabitants (the majority of the inhabitants) where damaged transport hinders them from visiting faraway healthcare centres, and this was especially difficult for women. NGO officials reported that during disasters, mothers could not leave their house for treatment. The damaged transport systems, responsibility for children and starvation for days prohibited them. In contrast, men were found to be leaving the house whenever they needed to. An NGO official who worked with Sidr victims, especially pregnant mothers and children, mentioned that '[a] man can walk to the pharmacy and get his medicines for a headache but a pregnant woman is not taken to the centre for a check-up. Three times check-ups are required during pregnancy but they cannot visit even once. If an NGO is not working in the locality, then women are not visiting the healthcare centres, their husbands do not even think of their wives'. So, medical preparations in the urban areas during the post-disaster periods were not helpful for rural women due to these additional social factors, gendered attitudes and traditions.

Again, medical teams, mostly male staff, reached the remote areas of Barguna three to four days after Cyclone Sidr. But during Cyclone Sidr, there was no female Government doctor appointed in Barguna, and the few female doctors who visited from the other districts mostly stayed at the urban centres. Here it should be mentioned that women in the field, whether they are a doctor, NGO officer or an inhabitant, all face the same cultural barriers. Insecurity, risk of physical abuse and humiliation all present additional challenges to her when trying to help disaster victims (Dominelli 2013), compared to male officials doing the same job. So, after disasters, mostly young male doctors were the ones to go to affected areas.

But their presence was not very helpful to the rural women. When young male doctors set up camp on open yards or at the school, young girls and women did not feel able to talk about their problems to them. The female patient's father, brother or husband might have taken them to the doctors but they could not express their problems properly to them. All these cultural issues became important barriers to the success of the medical response programmes.

Relief distribution

During Cyclone Sidr, vitamins and food supplements were distributed to children, pregnant and lactating women with other basic foodstuffs at the national level (Government 2008). One of the NGOs interviewed in Barguna distributed micro nutrient powder for malnourished children who were among Cyclone Sidr victims. Severely affected children were hospitalized and kept under supervision with the help of the Family Welfare Centre (FWC). During Cyclone Mahasen, the World Food Programme, through its partners Shushilan and Muslim Aid, distributed 12.6 MT of nutritious food for acutely malnourished children under five years and pregnant and lactating women among 2,291 cyclone affected households in Barguna and Bhola (DDM 2013). Analysis of Government reports and interviews reveals that the distribution of sanitary items and nutritious foods for women and their children got priority in the relief programmes of the previous cyclones in Bangladesh.

However, respondents in the present research had different opinions on this priority of giving relief to women. Respondents mentioned that, in reality, men were always first, though in many areas, women were also found to receive the relief. According to some of the Government officials, women took more medicine while the men were busy collecting relief, but NGO officials explained this in different ways. According to their experiences, women had been exploited in relief distribution in several ways. In most cases, women who were living near the municipalities received the medicine, and when men were busy collecting other relief. Women mostly collected the medicine when their family needed it, and this was a woman's additional family duty after disasters. Receiving relief also pleases the family when there is a great demand. However, women's own opinions on receiving relief from the roadside, standing in a long, crowded queue, have not been highlighted. For women to collect relief in a crowded place is not positively accepted by society, and women do not feel good standing in the relief lines. But most organizations prefer to include women as their beneficiaries. Again, the presence of women in relief collection is taken as a sign of success of the relief distribution programmes, which is highlighted by the media and responders. However, the question arises how much of the distributed items and money were consumed by the woman herself. A female NGO officer described her experiences, 'I feel women are being exploited during the disasters in regards of receiving relief. Women are receiving the relief but their relief supplies are mostly grabbed by their husbands. Very little remains for them. Only some aged women may have their own rights on relief. But unfortunately, women do not even know that they are being exploited'.

Again, the availability of proper food, enough in amount and frequency, is essential for pregnant women, but officials who worked at the field level mentioned that it is quite often pregnant women, especially in poor families, who could not get their food regularly, which becomes more difficult after disasters (Government 2008). An NGO official mentioned, 'After the recent cyclone Mahasen, there was rain for 15 days in Barguna which disrupted the daily life as well as the food availability and health of the pregnant mothers'.

Sanitation and women: humiliation and ridicule

During disasters, toilets get flooded, and this causes more problems for women. They have to share toilets with men, which is not socially acceptable. Although Government plans include the provision of separate toilets for women (MoDMR 2012), this was not implemented at many cyclone shelters. The consequences of the lack of separate toilets was not good for female victims. NGO officials mentioned, 'Women have to wait until night as they cannot go to the toilet in the day light whereas men can use toilets whenever they want'. Men are not affected by any social restrictions or traditions in this regard. The experiences of the female NGO officers show the extremes, 'We went there for distributing relief after Cyclone Sidr. We could not take a bath for eight days, could not eat rice for days and we were getting sick but we were not allowed to take a bath at the open pond. One day early in the morning we took bath and we have been criticised for that by the local inhabitants. We cannot forget those experiences of Cyclone Sidr'.

Now the question arises: when female officers face these humiliations, how can they be expected to take social risks for relief distributions in rural Bangladesh? And yet, if they do not take these risks, then the local rural women of remote parts of Bangladesh will face more severe consequences.

NGO workers also revealed that people teased the women for their dress when they had lost it during disasters. Women who were menstruating had to remain hidden for hours after Cyclone Sidr to hide their menstruation. During Cyclone Mahasen, men were observed laughing when they saw the sanitary napkins in relief packages. Because of local cultures, this is very humiliating for women and deters them from seeking help and relief. Because of these consequences, some NGOs in Barguna are now working on gender awareness, but is such training enough to change the mentality, or is it going to take more time? One NGO official raised a very important question in this regard, 'How can all of these sensitive problems of women be shared with those men who do not seem to care about women?'

Gender discrimination and violence reported by disaster managers

Gender discrimination in Barguna has been mentioned by the NGO officials during the interviews. They mentioned discrimination of different kinds (son priority, early marriage, negligence of women's rights) is experienced by most women in these communities but is worse for certain women. In some rural areas, young girls (14–15 years) are forced to get married to older fishermen. There is a trend revealed that these fishermen mostly leave their wives within a few years. When left by their husbands, young married girls are not accepted by anyone, not even by their parents, who are very poor. NGO officials mentioned, 'What a pity for a married woman who lives at her parent's house as an unwanted dependent'. These girls are the victims of social behaviour, traditions and culture prior to the cyclones hitting Bangladesh. During disasters, these women become the most vulnerable victims, facing the double hazard of the disaster and the social traditions. NGOs found they were unable to do anything for them because the local Governments and even the inhabitants forbid these girls from receiving relief. This view reinforces the position of these girls and young women as the most insecure and vulnerable in their own society. As Hossain (2008) writes, 'Human security is the most basic of public priorities and quite as vital as national security. But the frontline is not the border – it is in each community and every home' (Hossain 2008, p. 3).

According to the NGO officials, after disasters, human trafficking of women increases. Men from elsewhere visit the affected areas and target the young girls of the area. They convince them to leave the area to have a better life in Dhaka or others districts. Rape cases also increase after the disasters. Macdonald, Rashid and Pittaway et al. also mentioned about the increased incidence of verbal and physical harassment for women in the disaster-affected areas (Macdonald 2005; Rashid and Michaud 2000; Pittaway et al. 2007).

Families depending on fishing especially experience great problems during disasters. They are mostly the poorest, landless inhabitants of Barguna. If a trawler is lost during the disaster, 20–25 families become vulnerable. They do not have food reserves, even for the next day. Young girls of these families become extremely vulnerable. NGO workers found that most listed prostitutes of Barguna are from these most vulnerable areas and families. Similar findings were also mentioned in the research of Pittaway et al. 2007 on 2004 tsunami victims (Pittaway et al. 2007).

Working for women: disaster training and cultural barriers

Any project needs to consider local traditions to get successful results, especially in an area like Barguna, a remote coastal region of Bangladesh. NGO officials mentioned several aspects in regards of recruiting staff and volunteers in their projects. First of all, sending female staff to the affected areas requires special consideration and arrangements. Their safety and ability to access the vulnerable areas need to be considered top priorities. We can see harassment mentioned by women aid workers after the 2004 tsunami in Dominelli's research (Dominelli 2013). Secondly, they have to consider involving men in their women's projects to get successful results. An NGO officer described elaborately how they managed to continue their project,

> We have to include some men with female volunteers because during disasters a woman cannot cover ten houses alone. If she has any male company, especially husband or family members then she can. Again, our society is a

patriarchal society, if we want to improve the condition of women only, it will have a negative impact on us. We need men to participate in our programme to make it successful. If we go to any village they will not allow female groups to work with any NGO but if any one or two men could be involved from that village then it could be possible. We are still at this tradition. We could not get rid of this condition easily. We need more time. We also talk in accordance to religion. Men's presence and religious talk make people think well of the NGO. Now in our project area men help to continue the women's group to have meeting.

7.3 Plans and programmes for future disasters

The long-term aim of the Bangladesh Government's disaster management is to reduce the vulnerability to disasters of people, especially the poor, to a tolerable level. As a large number of people living below the poverty level are women, they are given priority in disaster management and distribution of relief goods. Gender equity is promoted through some programmes of the Ministry of Disaster Management and Relief that give women priority, not only as beneficiaries, but also involving them with implementing activities (Ministry of Finance 2014). The number of women involved at this level is not mentionable yet, e.g. in the Disaster Management and Relief Division, only 3.9% of the total number of employed officials are women (Ministry of Finance 2010). At the upazila relief office, there was no female officer employed in 2009–2010 (Ministry of Finance 2010, p. 17). Only one female member has been appointed among seven members in the Cyclone Shelter Management Committee at the local level (MoDMR 2012).

Beside these government initiatives, NGOs are also helping the government to improve conditions. At the local level, several programmes for women are taking place in Barguna after Cyclone Sidr. Although these programmes are currently being conducted on a small scale, they are inspiring and important initiatives which have the potential to improve the position of women in disasters in Barguna. For example, some programmes are mentioned in the following paragraphs.

With regard to disasters and healthcare, some NGOs in Barguna are working with pregnant women to train them to deal with disasters (pre-, during and post-disasters). The training focuses on how they are going to help themselves and their children. About 30,000 mothers are trained in one year. Women are also getting training on hygiene for six months with kits. In addition, one NGO has trained 300 female volunteers in every village of 15 villages (including Badarkhali) for the primary injury treatment and awareness. This training has also been received by the female volunteers of Tentulbaria village.

In regards to healthcare facilities, NGOs are now working on mother and child health with the Government to improve the FWCs by providing training to medical staff, nurses and medical equipments for safe delivery of babies and ambulance costs for transferring critical delivery cases to Barguna Hospital or Barisal. Though these NGOs are working in just a few specific health centres and only started last year, if these initiatives could be extended to all the healthcare centres of Barguna, then it would be a great help for the rural mothers to avoid unsafe home delivery. NGOs are also working on pregnant mothers' food security on a very small scale and with the village doctors to involve them with the MBBS (Bachelor of Medicine) and Diabetics Associations.

Again, several NGOs are training their beneficiaries and volunteers to help evacuees reach the cyclone shelters. One of the NGOs in Barguna is providing awareness training on signals and evacuation to the children in two local schools. They are trained to use toilets properly (male and female) at schools and cyclone shelters during the disasters. Now children are participating in planning and risk analysis for climate change. Again, NGOs are also working on adolescent sexual rights, prevention of early marriage and child abuse with the help of the local child groups and local Government. They have programmes of 'theatre for development' which have been very helpful to deliver the message to the local people by their children. NGOs are providing child counsellors for talking therapy at some FWCs. NGOs are also helping schools to continue operating after disasters.

We can hope that all these plans and programmes should be very helpful to accelerate women's advancement and strengthen their capacity to face disasters.

7.4 People's hopes and recommendations

Many opinions were also received during the research from the respondents about how to improve the conditions created during the pre- and post-disaster periods. As the inhabitants and primary disaster managers of the disaster-prone area, respondents wish for permanent solutions to the problems they report during the disasters. These include improving the local healthcare facilities, with a special focus on disasters; increasing awareness among the local administrators, medical staff and disaster managers; arranging special healthcare services for the post-disaster periods; and increasing safety for the inhabitants by building more cyclone shelters and embankments. As people with considerable local expertise about the problems faced in disasters, their suggestions are realistic and sensible and not necessarily all expensive. The suggestions mentioned by the respondents are described in the following paragraphs.

Improving healthcare services focusing on disasters

1   Respondents suggested in their interviews that local healthcare centres could be improved by posting more doctors and medical staff, confirming their presence at the centres during the disasters and storing essential medicines, especially those in high demand after disasters. More local medical staff and volunteers should be trained in primary healthcare to meet the needs of healthcare after disasters. Community clinics are mostly in a poor condition, and 10% are damaged. These should be improved so that remote areas get healthcare during disasters.

2   They also suggested the establishment of a new healthcare centre in their locality, which may be private, especially in vulnerable Tentulbaria village, to make sure that they will be able to get treatment after disasters. One

respondent mentioned in his interview that '[i]f any healthcare had been located in our village, it would have ensured the presence of medical staff after Cyclone Sidr, but we had to wait for the relief team which was really unpredictable . . .' (Male, age 28, rural area).

3   Building a temporary clinic or special medical camp after disasters in the locality, or in the cyclone shelters, would be helpful for receiving healthcare after disasters. As victims stay for a long time at the cyclone shelters, and all inhabitants know their locations, these temporary medical camps should be located at the cyclone shelters. Again, special emphasis should be placed on healthcare services at the local level for common health problems such as injuries and diarrhoea.

4   At present, injured people do not get proper attention during and after the disaster response programmes and no information could be found in regard of completing their injury treatments. Health impacts of disasters and accessibility to healthcare facilities should be recorded for better planning and response programmes.

Improving disaster management plans

1   Creation of a local disaster management committee that includes local house owners would be helpful to inspire them to contribute to evacuation plans. This might encourage a sense of responsibility to share their houses with evacuees during the warning periods.

2   All of the respondents (inhabitants, NGOs and government officials) made several suggestions in regards of cyclone shelters. The summary of these suggestions is as follows.

Cyclone shelters should be located within 2km of people's residences and at accessible locations. The number of shelters should be increased. Cyclone shelters should have a capacity of 500 people, with extra places for cattle and valuable belongings of evacuees. Again, people stay for a long time (five days) after cyclones, so they should have proper arrangements for cooking food, or they should receive supplied food in time.

3   There are still some vulnerable areas where there are no embankments. Embankments should be built to protect those areas, and some existing embankments need rebuilding.

Increasing awareness

Increasing awareness programmes was the most important suggestion for all respondents (inhabitants, NGOs and government officials). Social awareness should be increased to tackle political influence in healthcare sectors and create a good working environment for organizations, government and NGOs during disasters. Especially, government officials should take special steps and mention these in the emergency meeting of the disaster management committee (meetings called after the cyclone warnings are issued) to help the injured victims with

special care. Especially, doctors should be aware about the local response programmes so that they could be more prepared for the post-disaster periods.

Recommendations for women

Respondents also made several suggestions about the female inhabitants as the
most vulnerable group in cyclone disasters.

1 Special healthcare services for women should be arranged after disasters. To
   improve the situation, medical teams must reach remote areas after receiving
   warnings. They should stay there for two to seven days after disasters at the
   school, local houses, madrasha etc so that they can provide treatments just
   after the incidents. Government staff should be in the field with their emergency boxes. Again, temporary medical camps should be located at the union
   level with female doctors for three to six months in a protected place so that
   local women can visit there. Information about the availability of healthcare
   should be broadcast properly so that female victims know the location of the
   temporary healthcare services. Special sanitation and the sanitary needs of
   women should get priority during the disasters.
2 Female specialist doctors, gynaecologists and counsellors should be recruited
   in the district hospital or at the Upazila Health complex. A government official mentioned in the interview, '[A] female doctor is needed at every stage/
   everywhere. I requested this from the beginning but no female doctor wants
   to stay in Barguna, they want to stay in Dhaka'.
3 Data on pregnant women and new mothers should be regularly maintained.
   Special plans should be drawn up for transferring pregnant and sick woman
   to safe shelters before the disasters. Separate rooms should be provided for
   women in the cyclone shelters.
4 Creating a local woman's committee would be very helpful to get local
   women involved in disaster management. Local women should be trained
   on primary healthcare and receive necessary kits to be able to do their work.
   More female relief workers should be arranged. Again, every ward should
   have plans for male and female inhabitants. If local, educated people could
   receive training on primary healthcare, and they have a neighbourhood committee for disasters (especially with local female members), then Government officials could arrange a meeting with them before disasters and assure
   them that victims will have healthcare during disasters.
5 The amount of awareness training should be increased, with a focus on woman's preparation and the special needs of women in disasters.

7.5 Conclusion

This chapter has described the significant gaps between disaster management
plans and implementation in the remote coastal regions of Bangladesh. Several
environmental, institutional and behavioural factors (Figure 7.1) are combined

Environmental factors	Institutional factors
Severity of cyclone	Capacity of the organizations
Local cyclone risk	Presence of response plans
Location: Urban and rural	
Accessibility of the transport system	
Behavioural factors	**Socio-cultural factors**
Coordination difficulties	
Discrimination on the basis of political and socio-cultural status	Gender
	Cultural attitudes and norms
Corruption	
Professionalism and empathy	

Figure 7.1 Factors affecting response programmes during post-disaster periods

with socio-cultural factors to influence the response programmes and their successful implementation.

Despite preparations, response programmes depend on the severity of the cyclone disaster, the location of the affected areas and the accessibility of transportation system after disasters. Institutional capacity defines the limit, whereas behavioural factors, such as coordination among responders, presence of professionalism and corruption, influence the successful implementation. Lack of priority given to gender and culture in response plans creates a gap between plans and their implementation and discriminates among victims and deprives certain groups, especially women. Dominelli mentioned in her research that 'the impact of gender relations on women's lives is regularly ignored as a reflection of socially constructed realities that continue to be reproduced during disaster interventions' (Dominelli 2013, pp. 77–93).

Improvement of this situation – addressing the gap – is essential for the inhabitants of Barguna, and especially for marginalized groups, including women. Government and NGO officials and inhabitants have given their views and have made a series of recommendations which reveal the expectations of people who are living in the remote coastal region of Bangladesh, based on their first-hand experiences and observations of what happens before, during and after disasters. These findings and recommendations reveal the need to consider preparedness of the whole network of care and relief system, focusing on social relations, which cut across the objectives of aid operations.

References

Adams, A. M., Ahmed, T., Arifeen, S. E., Evans, T. G., Huda, T. and Reichenbach, L. (2013). "Innovation for universal health coverage in Bangladesh: A call to action." *The Lancet* 382(9910): 2104–2111.

Ahmed, F. (2007). "Bangladesh: Cyclone Sidr field report – BRAC." http://reliefweb.int/report/bangladesh/bangladesh-cyclone-sidr-field-report-brac-healthdirector-faruque-ahmed, accessed on 1/12/2012.

Ahsan, D. A. (2014). "Does natural disaster influence people's risk preference and trust? An experiment from cyclone prone coast of Bangladesh." *International Journal of Disaster Risk Reduction* 9: 48–57.

Alam, E. and Collins, A. E. (2010). "Cyclone disaster vulnerability and response experiences in coastal Bangladesh." *Disasters* 34(4): 931–954.

Balcik, B., Beamon, B. M., Krejci, C. C., Muramatsu, K. M. and Ramirez, M. (2010). "Coordination in humanitarian relief chains: Practices, challenges and opportunities." *International Journal of Production Economics* 126(1): 22–34.

BBS. (2006). *Population Census-2001, Community Series*. Zila, Barguna, Bangladesh Bureau of Statistics.

Berg, R. (2009). *Hurricane Ike tropical cyclone report*. Miami, Florida, National Hurricane Center.

Cash, R. A., Halder, S. R., Husain, M., Islam, M. S., Mallick, F. H., May, M. A., Rahman, M. and Rahman, M. A. (2014). "Reducing the health effect of natural hazards in Bangladesh." *The Lancet* 382(9910): 2094–2103.

CCC. (2009). *Climate Change, Gender and Vulnerable Groups in Bangladesh*. Dhaka, Climate Chamge Cell, DoE, MoEF, Component 4b, CDMP, MoFDM: 1–82.

CDMP. (2014). *Assessment Stakeholder's Role in Preparation For and Facing the Tropical Storm Mahasen*. Bangladesh, Ministry of Disaster Management and Relief.

Custers, P. (1992). "Cyclones in Bangladesh: A history of mismanagement." *Economic and Political Weekly* 27(7): 327–329.

Cyclone Preparedness Programme (CPP) and Bangladesh Red Crescent Society (BDRCS). (2007). "CPP at a glance." Dhaka. www.preventionweb.net.

DDM. (2013). *Disaster Preparedness Response and Recovery*. Bangladesh, Department of Disaster Management.

Debnath, S. (2007). "More shelters could save many lives." *Daily Star*, 24 November.

Djalali, A., Khankeh, H., Ohlen, G., Castren, M. and Kurland, L. (2011). "Facilitators and obstacles in pre-hospital medical response to earthquakes: A qualitative study." *Scandinavian Journal of Trauma, Resuscitation and Emergency Medicine* 19(1): 30.

Dominelli, L. (2013). "Gendering Climate Change: Implications for Debates, Policies and Practices." In Alston, M. and Whittenbury, K. (eds.) *Research, Action and Policy: Addressing the Gendered Impacts of Climate Change*. London, Springer: 77–93.

Fernandes, A. and Zaman, M. H. (2012). "The role of biomedical engineering in disaster management in resource-limited settings." *Bulletin of the World Health Organization* 90(8): 631–632.

Government. (2008). *Cyclone Sidr in Bangladesh: Damage, Loss and Needs Assessment For Disaster Recovery and Reconstructions*. Bangladesh, Government of Bangladesh.

Haider, M. Z. and Ahmed, M. F. (2014). "Multipurpose uses of cyclone shelters: Quest for shelter sustainability and community development." *International Journal of Disaster Risk Reduction* 9: 1–11.

Haque, U., Hashizume, M., Kolivras, K. N., Overgaard, H. J., Das, B. and Yamamoto, T. (2012). "Reduced death rates from cyclones in Bangladesh: What more needs to be done?" *Bulletin of the World Health Organization* 90(2): 150–156. doi:10.2471/BLT.11.088302

Hossain, M., Islam, M. T., Sakai, T. and Ishida, M. (2008). "Impact of tropical cyclones on rural infrastructures in Bangladesh." *Agricultural Engineering International: The CIGR EJournal* 2(X). www.cigrjournal.org, accessed on 5/2/2015.

Hossain, N. (2008). "The price we pay." *FORUM, a Monthly Publication of The Daily Star*, 3(1). www.the dailystar.net/forum/2008/january/price.htm, accessed on 22/9/2014.

Huque, R., Barkat, A. and Nazme, S. (2012). "Chapter 3: Public Health Expenditure: Equity, Efficacy and Universal Health Coverage." In S.M. Ahmed, T. G. Evans, A.M.

R. Chowdhury, S. Mahmud & A. Bhuiya (eds.) *Bangladesh Health Watch: Moving Towards Universal Health Coverage*. Dhaka, James P Grant School of Public Health, BRAC University: 25–32.

IFRC. (2010). "Bangladesh: Cyclone Sidr." Final Report, International Federation of Red Cross and Red Crescent Societies.

Islam, A. S., Bala, S. K., Hussain, M. A., Hussain, M. A. and Rahman, M. M. (2011). "Performance of coastal structures during Cyclone Sidr." *Natural Hazards Review* 12(3): 111–116.

Kunreuther, H., Meyer, R. and Michel-Kerjan, E. (2009). *Overcoming Decision Biases to Reduce Losses From Natural Catastrophes, Risk Management and Decision Processes Center*, USA, Risk Management and Decision Processes Center, The Wharton School, the University of Pennsylvania, 1–27.

Lai, T. I., Shih, F. Y., Chiang, W. C., Shen, S. T. and Chen, W. J. (2003). "Strategies of disaster response in the health care system for tropical cyclones: Experience following Typhoon Nari in Taipei City." *Academic Emergency Medicine* 10(10): 1109–1112.

Litman, T. (2006). "Lessons from Katrina and Rita: What major disasters can teach transportation planners." *Journal of Transportation Engineering* 132(1): 11–18.

MacDonald, R. (2005). "How women were affected by the tsunami: A perspective from Oxfam." *PLoS Medicine* 2(6): e178.

Mallick, B. J., Witte, S. M., Sarkar, R., Mahboob, A. S. and Vogt, J. (2009). "Local adaptation strategies of a coastal community during cyclone sidr and their vulnerability analysis for sustainable disaster mitigation planning in Bangladesh." *Journal of Bangladesh Institute of Planners* 2: 158–168.

Mallick, B. (2014). "Cyclone shelters and their locational suitability: An empirical analysis from coastal Bangladesh." *Disasters* 38(3): 654–671.

Ministry of Finance (2010). *Gender Budgeting Report 2010-2011*. Bangladesh, Ministry of Disaster Management and Relief, Ministry of Finance.

Ministry of Finance. (2014). "Gender budgeting report 2014–2015." Ministry of Disaster Management and Relief, Ministry of Finance. www.mof.gov.bd.

MoDMR. (2012). "Cyclone shelter construction, maintenance and management guideline 2011." Ministry of Disaster Management and Relief.

Murdock, D. (2005). "Multi-layered failures: Government responses to Katrina." *National Review*.

Murtaza, N., Shirin, M. and Alam, K. (2013). Enhancing disaster resilience in Borobogi Union, Barguna district, Bangladesh-Community Managed Reduction Program in Taltoli Upazilla, ACF.

Nahar, N., Blomstedt, Y., Wu, B., Kandarina, I., Trisnantoro, L. and Kinsman, J. (2014). "Increasing the provision of mental health care for vulnerable, disaster-affected people in Bangladesh." *BMC Public Health* 14(1): 1–9. www.biomedcentral.com/1471-2458/708, accessed on 17/12/2014.

Oloruntoba, R. (2010). "An analysis of the Cyclone Larry emergency relief chain: Some key success factors." *International Journal of Production Economics* 126(1): 85–101.

Paul, B. K. (2009). "Why relatively fewer people died? The case of Bangladesh's Cyclone Sidr." *Natural Hazards* 50(2): 289–304.

Paul, B. K. and Dutt, S. (2010). "Hazard warnings and responses to evacuation orders: The case of Bangladesh's Cyclone Sidr." *Geographical Review* 100(3): 336–355.

Pittaway, E., Bartolomei, L. and Rees, S. (2007). "Gendered dimensions of the 2004 tsunami and a potential social work response in post-disaster situations." *International Social Work* 50(3): 307–319.

Rashid, S. F. and Michaud, S. (2000). "Female adolescents and their sexuality: Notions of honour, shame, purity and pollution during the floods." *Disasters* 24(1): 54–70.

Ray-Bennett, N. S., Collins, A., Bhuiya, A., Edgeworth, R., Nahar, P. and Alamgir, F. (2010). "Exploring the meaning of health security for disaster resilience through people's perspectives in Bangladesh." *Health & Place* 16(3): 581–589.

Seaman, J. (1999). "Malnutrition in emergencies: How can we do better and where do the responsibilities lie?" *Disasters* 23(4): 306–315.

Shelter Coordination Group (SCG). (2008). "Barguna ward level shelter assessment." http://sheltercluster.org/sites/default/files/docs/BWLSA%20Barguna%20WardLevel%20Assessment%20Presentation.ppt, accessed on 25/1/2015.

Shultz, J. M., Russell, J. and Espinel, Z. (2005). "Epidemiology of tropical cyclones: The dynamics of disaster, disease, and development." *Epidemiologic Reviews* 27(1): 21–35.

Simatupang, T. M. et al. (2002). "The knowledge of coordination for supply chain integration." *Business Process Management Journal* 8(3): 289–308.

Stover, E. and Vinck, P. (2008). "Cyclone Nargis and the politics of relief and reconstruction aid in Burma (Myanmar)." *JAMA* 300(6): 729–731.

8   Conclusions and recommendations

8.1 Introduction

Cyclones are created by nature, but their impacts are intensified by human societies. This research has aimed at finding out the gender-specific health impacts of cyclones and identifying the factors shaping accessibility to healthcare in disasters. An in-depth investigation at the individual level using a qualitative methodology has been developed in order to focus on the complexity of gender relations and disasters in a very traditional society. A qualitative methodology is rarely used in investigating health impacts of disasters, but the present study has revealed that this methodological approach is very helpful in capturing victims' experiences, especially those of women. The findings make original contributions to disaster and gender studies, emphasizing the importance of gender analysis in disaster research as one way of helping to establish equality and equity in the health sector. Through its investigation, this research has argued that cultural attitudes, norms and traditions create and maintain differences in gender identity, roles and responsibilities, and strongly influence women's vulnerability and lack of healthcare access during disasters. Poverty also makes this situation worse. Now, this concluding chapter of the book will summarize all of the findings of the study and provide recommendations for improving the health conditions of vulnerable inhabitants, especially women, in disaster-prone regions.

Chapters 3, 4, 5 and 6 presented the research findings: the health impacts of disasters and the poor healthcare access during and after disasters, analyzing the influences of several inter-related factors. According to this study, environmental and behavioural factors are common contributors to healthcare inaccessibility in disasters, especially where socio-cultural factors like gender and poverty create highly significant differences among social groups. This thesis has argued that cultural attitudes, norms and traditions create and maintain differences in gender identity, roles and responsibilities and strongly influence women's vulnerability and lack of healthcare access during disasters. Poverty also makes this situation worse.

Again, Chapter 7 opens a new arena for the disaster managers and policy makers, presenting the factors affecting the implementation of prevailing disaster management plans. It identifies the importance of gender in disaster management plans and discusses recommendations from the local people and responders, as

Braithwaite and his colleagues (1994) advocate citizen participation for effective health promotions. The chapter describes the experiences of the responders and victims at the field level, and reveals the gaps, constraints and challenges in responding to emergencies and distributing relief during the post-disaster period. It also highlights that although disaster management has received much attention from scholars and authorities like governments and NGOs, gender-specific health impacts and healthcare access and their complex relationship with culture, tradition and social attitudes have not yet been fully recognized. Through its investigation, this research emphasizes gender-sensitive disaster management plans for successful implementation.

8.2 Health impacts of disasters

Natural disasters and their management are a great concern for the world. Disaster-prone countries like Bangladesh face a range of disasters almost every year (DDM 2013), and among these, cyclones are in aggregate the most severe and create the most casualties (EM-DAT 2015). Bangladesh has been working on cyclone early warning systems for decades and now has a well-prepared plan for warning dissemination, preparation and evacuation (Cash et al. 2014; Haque et al. 2012; Government 2008). However, the present research reveals that this evacuation plan did not work as expected either in Cyclone Sidr 2007 or in Cyclone Mahasen in 2013. It finds that several inter-related factors – environmental, demographic, behavioural, economic, socio-cultural and health factors – affect the situation, increasing the health impact of disasters.

For vulnerable inhabitants, making no preparations and taking shelter or staying in weak houses during cyclone attacks result in many different types of health impacts. Environmental factors, geographical location, availability of 'safe' shelter and behavioural factors, including lack of awareness and disbelief of warnings, delay evacuation, and all encourage people to leave their houses only once the cyclone is upon them. This delay is the major reason for injuries, death and psychological trauma. Again, remaining in weak houses also leads to health impacts, as flimsy structures are easily damaged by the strong winds and storm surges. Taking refuge at the cyclone shelter protects victims from injury but not from the psychological impacts of cyclones. This is because of the traumatic environment in the overcrowded shelters. This research has revealed many psychological impacts of disasters among victims. However, it is not easy to quantify these impacts because they are generally not recognized and indeed are commonly taken as the normal consequence of cyclones.

In addition to these environmental and behavioural factors, the health impacts of disasters are influenced by several other inter-related contextual strands, such as demographic, economic, socio-cultural and prior health status. Education and economic solvency help people to be more aware of and resilient to disasters, whereas poverty increases disaster vulnerability. Social factors like gender and its relation with culture, along with social attitudes and norms, have a strong influence on vulnerability, increasing a female victim's health problems in disasters, which will be elaborated in section 8.6 of this chapter.

8.3 Healthcare access after disaster: determinants and consequences

Healthcare access after disasters is also influenced by several factors (as reported in Chapters 5 and 6). Of these, environmental factors like disrupted transport and a lack of local healthcare facilities are most common for all victims. Most remote areas like Barguna do not have sufficient numbers of healthcare centres, treatment facilities and proper transport systems to carry patients. All these problems of healthcare services are intensified during the high demands of disasters. Rural inhabitants often have to depend on the distant healthcare facilities, which become more inaccessible in disasters. Again, medical responses tend to be concentrated in urban and easily accessible areas and reach the remote areas only two to three days after a cyclone. As a result, most injuries and several diseases of post-disaster periods remain untreated for a long time. Minor injuries might heal after prolonged pain and discomfort, but major injuries become worse and may turn into permanent disabilities. A few days after a cyclone, injured victims try to access medical treatment; however, not all of them can receive complete treatment. The present research has revealed that many Cyclone Sidr victims with major injuries are still suffering health problems, some even living with disabilities due to incomplete treatment. Actually, a victim's capacity to complete treatment is influenced by inter-related socio-cultural factors like gender and economic condition, which work in complex ways, create a complicated situation for each victim and result in impacts of different intensities. However, considering gender and its relation with other socio-economic factors, this research makes it clear that health impacts of disasters are worse for women than men. Women's healthcare access gets less priority, despite women facing more health problems in disasters.

8.4 Urban and rural experiences of cyclones

In the present research, Barguna municipality and Tentulbaria village were selected as typical of an urban and a remote rural area, respectively. The logic behind this is to reveal the differences between urban and rural areas experiencing disasters in relation to health and healthcare access. Comparing the conditions between the two areas, the present research reveals that the remote coastal location, lack of local healthcare facilities and poor socio-economic conditions create many similarities between urban and rural areas of Barguna in experiencing disasters, but with some differences.

Lack of shelters in disasters is a common problem for both urban and rural Barguna. Although Barguna municipality has higher number of buildings compared to Tentulbaria village, still there is a shortage of shelters in this area. Besides, problems with cyclone shelters, including an inaccessible location, vulnerability of the structure, lack of space for people and cattle and a filthy and unfriendly environment for women, children, older and disabled people, are common for both urban and rural areas.

Local healthcare facilities and transport systems are both important factors influencing healthcare access after disasters. In this regard, urban areas are in a better position compared to rural areas. Urban areas have more healthcare centres and better transport systems than the rural areas. These conditions help urban cyclone victims to receive treatment earlier than the rural areas. Relief aid is also distributed earlier to the urban areas compared to the rural areas. Disaster managers, responders, medical teams and volunteers mostly stay in and direct response plans from urban areas. Relief items are also stocked in urban areas, whereas it becomes impossible to carry relief items from urban areas to the distant villages due to damaged transport after disasters. These conditions reveal that better infrastructures increase resilience against disasters. However, the present research reveals an important finding: that there is similarity in women's healthcare access after disasters across urban and rural areas. Women's healthcare gets less priority despite the fact that they have more health problems. Women's residential locations do not make any difference in their healthcare access after disasters. Along with other factors, gender and its relation with social attitudes and culture are the most important influences on the healthcare access of women. So, an increase in social awareness is necessary to improve healthcare access, along with the improvement of the infrastructures in the rural areas.

8.5 Six years after Cyclone Sidr: were we ready for Cyclone Mahasen?

During Cyclone Mahasen, plans for warning dissemination, evacuation and response were improved in Barguna compared to those pre-existing Cyclone Sidr. Different national and local organizations, e.g. the Bangladesh government, NGOs and private organizations, were involved in these plans. Increases in awareness and making preparations were also noticed among the inhabitants. However, this improvement in awareness did not increase the number of evacuees significantly during the warning period of Cyclone Mahasen. Many of them planned to take shelter in a neighbour's house during the emergency, rather than at the cyclone shelters, due to the problems related to cyclone shelters. Compared to Cyclone Sidr, these problems of the cyclone shelters have not been resolved or improved within the six-year period. People stayed at their vulnerable houses or took risks to reach a safe shelter during Mahasen. These are the same phenomena that explain the higher number of injuries and deaths during Cyclone Sidr. So, the vulnerability of Barguna inhabitants is still very high and cannot be said to have changed much in the years following Cyclone Sidr.

The Bangladesh government now has a very good disaster management plan for cyclones, but still, several gaps and limitations have been revealed in implementing the plan at the field level (discussed in detail in Chapter 7). Similar to Cyclone Sidr, disturbed transport systems and a lack of different methods of transport became one of the major constraints for relief distribution after Cyclone Mahasen, despite plans for relief distribution being more developed and organized. The furthest remote areas received delayed relief or no relief at all. Medical teams could not reach remote areas quickly enough, while those areas still face an absence of healthcare centres in the locality. Actually, the lack of healthcare facilities is a major problem for the whole of Barguna district. These problems were a great challenge during Cyclone Sidr, and, after six years, the conditions have not improved. Barguna is thus still highly vulnerable to any cyclone in regards to healthcare access after disasters.

Again, one of the significant factors affecting healthcare access is gender, which still has not received proper attention in disaster management plans. Although the plans advise giving priority to women in evacuation, in getting space in cyclone shelters and in receiving relief aid, the practical implementation of these plans is very difficult. This difficulty is mainly created by the lack of consideration to gender in relation to other social factors. During Cyclone Sidr, these problems were present, and after six years, we can see that conditions had not improved in Cyclone Mahasen. This could have led to a large number of victims if Cyclone Mahasen had turned into a more severe cyclone. The inhabitants of Barguna, especially women, are still highly vulnerable to cyclones in regards to health impacts of disasters. Culture, traditions and social attitudes create differences in gender relations and responsibilities, increasing the vulnerability of women. These conditions need special attention. The strong relation of gender and health impacts and its importance in the disaster management plans are discussed in the following sections.

8.6 Gender, the most significant factor

In the present research, gender is found to be the most prominent factor of all. 'Women' is a status which amplifies suffering and is associated with more injuries, greater health impacts and less healthcare access during disasters, with few exceptions. Previous chapters presented evidence of this from local stories and NGOs and other professionals that, compared to men, women are considerably more vulnerable in disasters in coastal Bangladesh (Begum 1993; Fordham 1998; Nasreen 2004; Rahman 2013). Social attitudes and culture create these differences among the members of the society (Weisman 1997; Bari 1998; Nahar et al. 2014) and make the situation worse for women, especially those in poorer households. Within all groups, whether rich or poor, educated or illiterate, earning or non-earning, single-headed families, people with disabilities and so on, female members are more vulnerable.

Higher impacts of disasters on female victims have also been revealed in other research (detailed in Chapter 1). Neumayer and Plumper analyzed a sample of 141 countries over the period from 1981 to 2002 and concluded that 'it is the socially constructed gender-specific vulnerability of females built into everyday socio-economic patterns that leads to the relatively higher female disaster mortality rates compared to men' (Neumayer and Plümper 2007, p. 551). Greater impacts of disasters on women have been revealed generally, for example, in the 2003 heat wave in Europe, where significantly more elderly women died than men among 20,000 people; the cyclone in 1991 in Bangladesh (CCC 2009); the 2004 tsunami in Ache, where 75% of those who died were women (Kandaswany 2005 in CCC 2009); and there were also greater health impacts on women in the recent UK floods (Tunstall et al. 2006).

However, it is a matter of concern that after all of these documented experiences from previous disasters and interventions to reduce the health impacts of disaster, gender is still the most significant factor shaping women's higher vulnerability. Ikeda (Ikeda 1995) in her research on the 1991 cyclone in Bangladesh mentioned that gender was one of the most vital factors explaining the higher casualty rate for women; and yet, 20 years after her research, conditions have not changed. The number of casualties may have fallen (in the 1991 cyclone, 90% of the victims were women and children, and in Cyclone Sidr in 2007, the male-to-female death ratio was 1:5, Ahmed 2011), but similar contexts and reasons for gendered vulnerability and disaster health impacts have been revealed among Barguna women during Cyclone Sidr and Cyclone Mahasen. Rahman (2013) also mentioned in his research that 'women are more vulnerable to climate disasters than men through their socially constructed roles and responsibilities, and their relatively poorer and more economically vulnerable position, especially in the developing world like Bangladesh' (Rahman 2013, p. 81).

Now, a more elaborate explanation is needed in order to know why and how cultural attitudes, beliefs and norms influence women's vulnerability and healthcare access in disasters, as these are inter-related with other behavioural and economic factors.

Women and health impacts of disasters

To know the full picture, analysis is needed of conditions prior to a cyclone attack. If we consider the conditions of women living in disaster-prone Barguna, it can be seen that gendered attitudes within society do not give opportunities for females to be more self-protective and self-dependant when disaster strikes. They do not have proper access to essential disaster information and important knowledge about the locality. Sultana mentioned in her research that 'notions of honour and shame are often invoked to control gendered mobility, and such issues are internalized and practised by both men and women in reinforcing who is to go where, for how long, how far away, and why' (Sultana 2011, p. 299), and this practice creates a sharp difference between males and females in receiving information during disasters. Women have to depend on males for disaster information as well as decisions for evacuation. Besides, taking responsibility for children and household goods during disasters makes them very anxious because local areas are not familiar to them; they feel physically unfit in strong winds and water and face a lack of life-saving skills (swimming, climbing trees). Their common dress, the 'Sari', and long hair become a 'death noose' during the cyclones, being aggravated by gender-biased social attitudes towards women's dress compared to men's dress.

All these problems are intensified for poorer women. Like other work (Cannon 2002; CCC 2009; Sultana 2010; Rahman 2013; Nahar et al. 2014), this present research found a strong relation between poverty and women's vulnerability. Cannon (2002) stated that '[v]ulnerability to hazards involves a complex interaction between poverty and gender relations, in which women are likely to experience higher levels of vulnerability than men' (Cannon 2002, p. 48), and this is also reflected in Sultana's findings in her research on Bangladesh (Sultana 2011). Poor women's vulnerability starts from the poor state of their houses and vulnerable location, which make them take refuge in cyclone shelters that are mostly unfriendly to women. Most of the cyclone shelters do not have separate spaces and privacy for women, especially for priority groups (pregnant and lactating and sick women), increasing the suffering of these women, even within the separate women's spaces. Besides, damage to toilets during disasters brings additional health impacts for women because of social attitudes towards women's sanitation needs. Women are discriminated against compared to men in using toilets and bathing.

During the post-disaster periods, socio-economic conditions are also found to be one of the major determinants influencing surviving capacity and resilience of the inhabitants. More families that are solvent are able to store some food and water in their comparatively stronger houses for the post-disaster difficult periods. Their safety net is stronger, whereas poor inhabitants have to leave their houses and depend on response programmes, relief distribution and recovery programmes. However, as we have seen, this is problematic for poor women because of social attitudes (elaborately discussed in Chapter 6). 'At a poor household, surviving through a cyclone turns into a nightmare for the woman/women in the aftermath of the event. The hazards easily become a disaster to a poor woman living in cyclone prone areas' (CCC 2009, p. 34).

These socio-cultural factors also amplify psychological impacts among the victims, especially among women, besides the traumatic cyclone itself. Though psychological problems are not identified by medical personnel, descriptions of their feelings during interviews suggests that women are more vulnerable to the psychological impacts of cyclones compared to men because of socio-cultural factors. Similar findings were also revealed in research on the health effects of flooding among residents in England and Wales in 30 locations (Tunstall et al. 2006). Again, Nahar and colleagues (2014), in their research on Bangladeshi disaster-affected people, also mentioned 'a large proportion of the survivors – and especially women and the poor – are already traumatized and need psychological support' (Nahar et al. 2014, p. 8).

Women and healthcare access

It is a matter of concern that, despite all these psychological and physical health impacts, women's healthcare needs did not get proper attention but rather were subject to the 'inverse care law' (Alam and Rahman 2014, p. 68), where more injuries received less care. Males are found to be in better position to access the minimal healthcare services available after disasters. Besides the common determinants (availability of healthcare facilities and accessibility of transport systems) of healthcare access, female victims face socio-cultural, economic, behavioural and health factors which strongly influence their access to proper medical treatment after disasters.

Dependency of women and healthcare access

Women, especially those living in remote areas, are dependent on men at every step of receiving treatment for their injuries after disasters. They need a male chaperone and family decision to receive healthcare. However, men are mostly very busy repairing damage or collecting relief, coupled with their lack of awareness and negligence of female's special healthcare needs. Again, family decisions are influenced by the 'son priority' and economic conditions. The family is a strong institution in Barguna society, where gendered politics are cultured in a patriarchal environment. Mainly, adult and senior male members and sometimes the senior female members (mothers, mothers-in-law) are the decision-makers in the family, and they take decisions for the young women and junior members.

Poverty and women's healthcare access

Getting injury treatment becomes very difficult after disasters, and a lack of money determines the type of treatment, healthcare centres and especially the timing of receiving treatment. Poor, injured victims often ignore their injuries, either due to lack of awareness of their seriousness or shortage of money, and sometimes because of prioritizing collection of relief donations rather than getting treatment. Delays and incomplete treatment can be a reason for injuries becoming lifelong disabilities for some of the poor victims. Poor, injured women face insolvency in a more intensified form, being dependant on family money (Momsen 2010). Similar findings were revealed in a research on coastal poor residences in Bangladesh: 'Women do not even dare to think of affording health care, they just share it with someone trustworthy and accept any consequence' (CCC 2009, p. 49).

Women's personal approaches and behaviour

Another important factor, women's personal approaches, also plays a significant part in seeking healthcare after cyclones (Paolisso and Leslie 1995). The personal choice to delay treatment is found more among women because of conservativeness, shyness, lack of awareness, consideration of the cost of treatment and prioritizing household responsibilities on top of their own healthcare needs.

Men and impacts of disasters on health and healthcare access

Considering 'men' and their vulnerability shows that gendered behaviours within both family and society help men to enjoy more freedom, autonomy and resilience in disasters, where they are not fragmented by socio-economic and political status. In some cases, however, gender roles and social expectations bring vulnerability for men in disasters, which is noted as the 'hidden impacts on men' by Fordham (Fordham 1998, p. 127) in her research on flood victims in Scotland. During the life-threatening conditions of cyclone attacks, local men are the main volunteers helping others, taking responsibility for the whole family and its valuable belongings, which is a major reason for men's deaths and injuries during disasters. However, gender works differently for them as men are taking on these gender roles intentionally. They can leave or loosen their gender role and responsibilities whenever necessary to save their own lives. Mostly, for them, society is found to be more relaxed in accounting for their actions and activities, whereas women are under pressure to save their lives during disasters, maintain their responsibilities and 'honour', or they will be humiliated, dishonoured, neglected and even charged in the family and society after disasters. Although the cultural context is very different, this type of imposed responsibility and increased vulnerability of women has also been noticed among the flood victims of Scotland mentioned by Fordham in her research (Fordham 1998). In regards to healthcare accessibility during the disasters, men's access is much greater than women's. If they can arrange money, time and transport, they visit the healthcare centres. Men's treatment gets delayed only when they do not identify the seriousness of their injuries and are busy collecting relief aid.

The theorization of complex gendered relations

In summary, it can be said that both men and women are confined by gender identities, roles and responsibilities, becoming the victims of disasters 'in their attempts to survive in hazardous environments' (GTZ 2005 in CCC 2008; Sultana 2010, p. 43). However, it is not straightforward to theorize gender identities, responsibilities and gendered relations in disasters. Gender is a complex concept, centring upon the fundamental idea of 'what it means to be defined as man or woman has a history' (Rose 2010, p. 2). Before the last decades of the twentieth century, it was popularly believed that gender differences were based in nature and these 'natural differences', and the relations between men and women were not questioned (Rose, 2010, p. 3). Gender was originally used by feminist scholars 'to mean the cultural construction of sex difference, in contrast to the term "sex", which was thought to mean "natural" or "biological" differences', but feminist scholars have since identified several problems with the sex/gender distinction (Rose 2010, p. 3). The philosopher Judith Butler has elaborated a way to understand sex and gender, arguing that 'sex is cultural achievement with bodily (material) consequences' (Rose 2010, p. 20). Butler (1990) defines gender as follows: 'when the relevant "culture" that "constructs" gender is understood in terms of such a law or set of laws, then it seems that gender is as determined and fixed as it was under the biology-is-destiny formulation. In such a case, not biology, but culture, becomes destiny' (Butler 1990, p. 11). Butler also adds that 'the body itself becomes gendered through repeated bodily acts', a process she terms 'performativity' (Rose 2010, p. 20). The sociologist Raewyn Connell (1987) also argues that 'gender becomes incorporated into the body in practice – in acting and interacting in the social world' (Rose 2010, p. 20).

However, problems arise in differentiating and defining the social roles of genders, especially in disastrous conditions. According to Nicholson (1994), gender is assumed to have a clearly understood meaning; however, it is actually used in somewhat contradictory ways, so that it is not easy to determine 'whose definition of women or women's interest' is in play (Nicholson 1994, p. 103). In the present research, it has been noticed that a strongly traditional society shapes 'personality and behaviour, it also shapes the ways in which the body appears' (Nicholson, 1994, p. 79). This patriarchal society expects masculine and feminine traits in men and women respectively. However, not all members of society feel the same. 'Masculinity is a reaction against passivity and powerlessness, and with it comes a repression of a vast range of human desires and possibilities: those that are associated with femininity' (Kaufman 1987, p. 486). Evidence from the present research shows that men do not, cannot or do not want to perform all of the expected masculine roles (e.g. be stronger all the time, rescuing victims, taking responsibility for all family members and valuable goods in cyclones). Equally, women do not, cannot or do not want to perform all the expected feminine roles, such as being dependent all the time. As I have shown earlier in this thesis, and as theories of intersectionality suggest, gender relations intersect with many interrelated variables to deepen inequalities between men and women and increase complexities in gender roles. Both 'men' and 'women' performed complex and mixed gender roles during disasters; for example, some men left their family to save themselves or were rescued by others, and women became stronger, rescuing others and forgetting their expected dependant social role. At other times, too, performing or wanting to perform the other's gender role was evident among both men and women. Some older women were found to be involved in practising the patriarchal dividend, and some men were found to be keen to change gender roles and responsibilities, especially wanting women to be independent and making their own decisions. Butler (1990) also mentioned this presence of traditionally 'feminine' attributes in men, and as she argues, 'it is no longer possible to subordinate dissonant gendered features as so many secondary and accidental characteristics of a gender ontology that is fundamentally intact' (Butler 1990, p. 33). Kaufman (1987) also explained masculinity as 'terrifyingly fragile because it does not really exist in the sense we are led to think it exists; that is, as a biological reality – something real that we have inside ourselves. It exists as ideology; it exists as scripted behaviour; it exists within gendered relationships' (Kaufman 1987, p. 487). But this 'dissonant play of gender attributes' raises questions, creates confusion and complexity and 'fails to conform to sequential or causal models of intelligibility' to understand gender as substance and 'the viability of man and woman as nouns' (Butler 1990, p. 33). Nicholson (1994, p. 80) also comments that this complexity 'generates obstacles in our abilities to theorize' the genders and understand the differences between and among them which have been discussed in the present research.

Intersectionality, gender and culture

Notwithstanding this complex theorization of gender identities, responsibilities and gendered relations in disasters, this research has clearly demonstrated that socially gendered expectations make women dependant and vulnerable in everyday life (Fordham 1998) and also shown that during disasters, women make up a disproportionately higher number of victims. Again, women's healthcare gets less priority (Cannon 2002; Sultana 2010; Alam and Rahman 2014; Nahar et al. 2014) and gets delayed, despite women facing more health problems. This is amplified by the 'gendered crisis' (Sultana 2010, p. 48), poverty and increases in psychological stress due to intensification of 'the sense of powerlessness and marginalization' during the critical times of disasters (Hossain et al. 1992 in Sultana 2010, p. 48). Women as a group 'generally face greater marginalization and oppression than their male counterparts' (Sultana 2010, p. 44).

However, it was expected that this research would require an intersectional analysis, which 'foregrounds a richer and more complex ontology than approaches that attempt to reduce people to one category at a time' (Phoenix and Pattynama 2006, p. 187) and thus should help to understand the complexity of gender relations with various interacting social variables. I expected that I would find in Barguna that 'women are not a homogeneous group, because intersectionality with class, caste, religion, age etc affects the resources, rights and responsibilities that any woman has' (Sultana 2010, p. 44) and that these create 'certain patterns' of disaster impacts (Sultana 2010, p. 46). However, although there are indeed these differences between women in Barguna (elaborately discussed in Chapters 4 and 6), the present research also emphasizes that remote location and strong cultural attitudes, beliefs, norms and traditions around gender create homogeneity among women from different social groups in the pre, during and post periods of overwhelming disasters like cyclones. Such is the powerful influence of gender cultures on women's lives. Considering cross-sectional factors like age, education, economic conditions and employment shows that these statuses increase women's physical protection and resilience (stronger houses in safer locations) and improve awareness of health and healthcare, but do not always assure healthcare accessibility after disasters. These statuses help women to participate in family discussions, but their opinions do not always get proper attention and increase their empowerment and values in the family as well as in society. Mostly, families (adult men and, sometimes, senior female members) take the final decisions on women's healthcare access (The Prothom Alo 2015a). Employed women do not always have control over their own earnings and resources to meet their necessities like healthcare in disasters, just as is the case in normal times (Rezwana 2013). Rather, 'little disposable income remains for use by women for their own health needs' (Paolisso and Leslie 1995, p. 61). Similar findings have been reported in the *National Newspaper* in Bangladesh in a roundtable discussion on the occasion of Women's Day on 8 March 2015. It was reported that development of Bangladeshi women has been achieved during the last two decades, but the empowerment of women still needs to be increased from the family level to national level. Women have the least control of their resources, their decision-making opportunities are too narrow and they are hugely victimized (The Prothom Alo 2015b).

In the present research, utilizing the concept of intersectionality helped to understand that an 'individual's social location as reflected in intersecting identities' (Shields 2008, p. 301) is complexly inter-related with gender identities and roles. However, according to Butler (1990), 'there is no gender identity behind the expressions of gender; that identity is performatively constituted by the very "expressions" that are said to be its result' (Butler 1990, p. 34). Gender solidifies into a form by repeated stylisation of the body and acts over time. The present research reveals that inhabitants of the patriarchal Barguna society reiterate gender in this way; they can be seen to perform masculinity and femininity and follow certain roles and responsibilities produced by socio-cultural processes. These roles and responsibilities, in relation with several social variables, produce inequalities between men and women. Despite some inhabitants, both men and women, wanting to change social attitudes and practices around gender, an inter-sectional analysis highlights the complexity in doing this. Many inter-related variables and complex relations have to be changed, if gender relations are to made more egalitarian. A number of forms of oppression must be identified if people are to become aware of their participation in oppressions. In addition, men (and some women) should become allies.

Some changes and future hope

Here it should be mentioned that a little difference has been noticed among women and some men in the present research. They identified the need for women's development and empowerment and mentioned their wish to change the gendered attitudes of society, but as members of traditional families, it is 'difficult for men to recognize alternatives or to understand women's experiences' (Connell 2003, p. 9). Their gender identity, roles and responsibilities confined by the local culture and traditions create a difficult situation for them to be active in social change. However, this consciousness might be one of the reasons for higher girls' attendance at primary schools and the increasing number of working women who can enjoy a little empowerment, taking part in the family decisions. Again, 'Purdah' is no longer a constraint on women visiting a male doctor in an emergency, though female doctors are still highly preferred. It is matter of hope that, though not at any measurable rate, at least some families are becoming aware about their daughters and their wellbeing, even living in a very traditional, patriarchal society.

The above discussions show the detailed aspects of culture and its relation with gender, creating the larger patterns among the people. These findings reveal the importance and seriousness of considering culture in defining gender and vulnerability and their strong influence on women's lives, especially in poorer economic backgrounds. They also lead to further thoughts, suggesting a need to rethink prevailing disaster management plans to consider gender and culture. This will be discussed in the following sections, taking into account the present research findings.

8.7 Comprehensive, gender-sensitive disaster management plans and recommendations

The present research has revealed the limitations and gaps in prevailing disaster management plans and policies in light of the challenges created during disasters. It shows that despite its importance, gender is not considered as an inter-related factor with social and cultural attitudes, norms and traditions in the prevailing disaster management plans and policies (discussed in Chapter 7). Analyzing the research findings and prevailing development approaches, this study emphasizes the need for disaster management plans to be comprehensive (Alam and Rahman 2014), gender-sensitive (Fisher 2010) and to focus on 'gender mainstreaming' (Momsen 2010, p. 15) and 'WCD' approaches, thus highlighting women's agency in the foreground, side by side with culture, society and economic influences (Bhavnani et al. 2003) (more in Chapter 1). Gender equity (Sen 2002; Doyal 2000) and equality should be a special concern for these plans, creating a supportive environment for all. Dominelli mentioned in her research that 'making the differentiated experience of women, men and children explicit in disaster intervention process is essential if all voices are to be heard' (Dominelli 2013, pp. 77–93). Again, in this process, participants are expected to be supportive of each other and become allies 'to contribute to change' (Bishop 2002, p. 110), rather than continuing exploitation through power politics at the family and social levels. However, it should be remembered that these plans cannot impose new ideas because these require massive changes in the existing culture, traditions, beliefs and norms, potentially leading to chaos and the rejection of plans by the local society. Rather, there should be a focus on those social attitudes which create the gendered difference and discrimination in society. So, disaster management plans should be more sensitive and more concerned about all of these social aspects. This conclusion supports Dominelli's suggestion that 'social work educators and practitioners from overseas have to interrogate their internationalism to ensure that they do not damage the people they aim to help further, but work with them in local empowering partnerships that they control' (Dominelli 2014, p. 258). Following the findings and suggestions of the present research, the following recommendations aim to improve prevailing disaster management plans taking examples from the present study. These fall into five categories.

Integrating and prioritizing gender and socio-cultural practices, norms and beliefs in policy and practice

Integration of gender and socio-cultural issues is challenging but very important for the improvement of the prevailing disaster management plans and policies of disaster-prone countries like Bangladesh (Enarson 2002; Fordham 2003, Parvin et al. 2008; Alam and Collins 2010, Ray-Bennett et al. 2010; Sultana 2010). Several socio-cultural issues have been raised regarding previous cyclones which became a great challenge to the success of the prevailing plans. Addressing these issues, particular recommendations are made in the following paragraphs.

Priority groups and evacuation plans

Cyclone warning forecast systems in Bangladesh are very effective in alerting coastal inhabitants. However, problems arise when inhabitants do not have cyclone shelters near their residences. Walking a long distance also becomes a great problem for women, pregnant, sick and older people, children and people with disabilities (Huq-Hossain et al. 2013). They need support. Though national Government plans state an intention to help women, children and people with disabilities to reach the cyclone shelters on a priority basis, it is not possible for the local Government in most of the vulnerable areas to actually do this due to lack of money, vehicles and manpower. To improve conditions, a proper plan is needed. For instance, local Governments could at least arrange a few vehicles to carry the most vulnerable people on a priority basis. If a proper priority list (including children, pregnant women, new mothers, older and disabled persons) and plans can be made prior to cyclones, evacuation of the vulnerable groups could be completed properly in several shifts. For example, among the women's group, pregnant women and new mothers, sick and old women should be enlisted first among all women in the priority list. Local Government and NGOs should make these priority lists and keep them updated.

Again, making a priority list will be helpful to increase the awareness about vulnerable groups, who still do not get priority within family and society. These groups should be introduced to the local inhabitants by arranging awareness programmes and training so that they and their families will be well prepared for a cyclone evacuation. Besides, the priority list will assure the vulnerable group that they will be taken care of, which will also reduce the mental stress associated with evacuation. Efficient use of space within the cyclone shelters would also be very helpful using a priority list.

Providing food and healthcare for evacuees before cyclone attacks

Availability of sufficient food during the warning period is a great problem in cyclone shelters. The cyclone shelter management committee and local NGOs should have plans to provide food to the evacuees, especially children, the elderly, the sick and pregnant women from the beginning of the evacuation and for a few days after the cyclones. Besides, all cyclone shelters should have a first aid box with essential medicines. Many people get injured on the way to the shelters, and many sick people take protection in shelters. Even cases of childbirth in the cyclone shelter have been reported. Therefore, management committees could appoint volunteers (both male and female) with first aid training for each shelter. This will save many injuries from worsening before people have the chance to go to a medical centre.

Planning for medical response in relation to geographical locations and socio-cultural demands

Ensuring medical support after the cyclones is very important to reduce negative health impacts. Search and rescue immediately after cyclones and early care of injuries in the field would be an effective response plan to reduce the health impacts of disasters (Fuse and Yokota 2012). The Bangladesh Government and NGOs have specific medical response plans. According to these plans, additional numbers of medical teams are deployed, and local healthcare centres are kept ready to meet increased demand after disasters. However, this research reveals that medical teams from the urban areas cannot reach the remote rural areas for two or three days due to disrupted transport after the cyclones, which raises questions about the success of the response programmes. Therefore, plans could be made for medical teams to arrive in the remote areas and to stock enough medicines at the local level before a cyclone emergency (Government 2008). They can be based at the cyclone shelter or local school or madrasha and continue providing the treatment facilities for several days after the cyclone. The length of stay of the medical teams should also be long enough to provide full treatment.

Another recommendation should be made in regard to medical team members. At present, medical teams mostly consist of male doctors, but a traditional society such as that in Barguna needs female doctors and staff. Government and local private health organizations should recruit more female doctors so that they are able to help female victims after disasters. Paolisso and Leslie 1995 mentioned in their research that '[h]ealth interventions must take into consideration the important characteristics of women's lives that affect their ability to address these problems' (Paolisso and Leslie 1995, p. 55). Medical teams should also choose a closed place for providing treatment facilities, for instance, rooms in the cyclone shelters, Government offices and schools, for the privacy of patients. Highlighting these social needs in disaster management policy would be helpful to increase female healthcare accessibility.

Again, choosing the relief items with regard to cultural practice and traditions is necessary to meet demand after disasters. Among the relief items, sanitary items are included to provide facilities for the local women, but it has been revealed that the demands are different culturally, local women needing long clothes. These situations reflect other research on tsunami victims by MacDonald (2005) and Begum (1993) on the 1991 cyclone in Bangladesh. Government, NGOs and other organizations involved in relief distribution should be more aware in this regard.

Incomplete treatment and help

Saving injured, poor people from the consequences of incomplete treatment should be important in the disaster management plans. However, the present research has revealed the situation to be otherwise. So, steps should be taken to provide any incentives, money or donations, especially for injured, poor women. Especially, volunteers should receive proper tools and medical kits and assurance of medical support for injuries. Private healthcare centres of Bangladesh could take initiatives in these regards.

Involving more women in disaster management committees

The presence of more women at the decision-making level of disaster management would be helpful to highlight the problems of local women during disasters. Dominelli suggested in her research that 'women should be involved in these activities from the word "go" and throughout, including in their intervention' (Dominelli 2013, pp. 77–93). Although the Bangladesh Government claims it is concerned about improving the gender balance by involving more female officials in disaster management, statistics from recent years tell us otherwise. Again, the presence of few women planners, managers and officials creates imbalance, and so women's voices are not represented properly. For example, the presence of one female member in a cyclone shelter committee will not be that influential at the local level of a traditional society like Barguna. So, employing and involving more female members in the government's and NGO's disaster management committees is recommended to improve the situation.

Improving local disaster management committees and the availability of essential information

The lack of a budget and resources is common for local committees in developing countries, but the implementation of disaster management plans at the field level is highly dependent on this committee. To solve the problems, more disaster management committees should be formed in the smallest administrative units using local resources and involving more volunteers from the locality. For example, the UDMC works within the smallest administrative unit of Bangladesh, facing several shortcomings, such as lack of funds, manpower, information and resources. To help the UDMC with more manpower, village disaster management committees (VDMC) could be created to take the disaster management plan to the micro level (Murtaza et al. 2013). Involvement of more local inhabitants in the VDMC would be helpful to get connections with the local inhabitants. Again, creating a local 'Women's Forum' (Murtaza et al. 2013) and including women in the VDMC would be very helpful to engage more women in the decision-making level of disaster management.

To make an effective emergency response plan, information on morbidity should be available immediately at the local level to quantify the healthcare needs after the disasters (Fernandes and Zaman 2012; Cash et al. 2014; DDM 2013). The government should be more concerned about this. Besides, information on socio-economic conditions of the inhabitants, updated information on the people from priority groups (Huq-Hossain et al. 2013) and local healthcare centres and available healthcare facilities are also essential. For example, during the recent Cyclone Mahasen, the lack of this information was a constraint for the responders; it meant delays in response programmes and lowered their effectiveness.

Therefore, a plan could be made for the availability of this essential information after disasters. For this purpose, a 'common information hub' (DDM 2013, p. 50) could also be established, as proposed by the government in their annual report on disaster preparedness, response and recovery in 2013. Again, healthcare information could also be arranged in an 'electronic health information system' (Adams et al. 2013, p. 2109) to improve the transparent and effective health coverage among the inhabitants after disasters. All this information could be available to the UDMC and VDMC and local inhabitants from the Union Digital centres. The Bangladesh government has been implementing an 'Access to Information (a2i)' programme since 2007. Under this programme, Union Digital centres were inaugurated at the union level from 2010 (a2i 2013). These centres at the union level would be a very good source of information and helpful for national and foreign responders to plan for emergency responses after disasters (Begum 1993).

Mobile phones are commonly used in Bangladesh. They are potentially a good source of information and could be used to assess local health status and needs

(Cash et al. 2014) and disseminate awareness about post-disaster health impacts among vulnerable groups.

Increasing awareness of gender and impacts of disasters

Increasing awareness of gender differences, discrimination and gendered attitudes in society will be very helpful to solve many of the problems women face, both in everyday life and in disasters. For these major changes to take place, however, alliances are required between different groups in society: men, women and rich and powerful people. Making progress on the issues this thesis has identified will involve getting allies on side: people who 'take responsibility for helping to solve problems of historical injustice' (Bishop 2002, p. 110).

The present research suggests that men mainly belong to privileged groups in Barguna; however, women in older age groups or from wealthier socio-economic groups also participate in different forms of oppression. According to Bishop (2002), 'an experience of oppression is necessary for a person to learn to be an oppressor' (Bishop 2002, p. 111). She explains that most people in society do not know about these connections between different forms of oppression and participation in oppression, or becoming an ally in a given setting. But becoming an ally needs both a clear understanding of oppression and an identification of one's own position as an oppressor or oppressed. So awareness training is necessary to educate and inspire inhabitants to become allies. Most women, and some men in the present research, expressed their wish to become active in changing social attitudes towards gender, but they do not know how, and they are also concerned about being excluded from the collective or humiliated among their peer groups. So, the Government, NGOs, educational institutes and civil society organizations should come forward to support these potential allies, for example, by arranging awareness programmes and training on gender differences, discrimination and gendered attitudes of society. Some recommendations for such programmes will be made in the following sections.

Women's empowerment, capacities and strength

Awareness programmes could be arranged by the Government, NGOs and civil societies on gender equality and equity relating to their culture, traditions and religions and focus on women's empowerment, strength and capacity (Enarson et al. 2007). People of different social groups and ages should be the participants of these programmes. This would have broad social effects (Connell 2003). Bari in 1998 also emphasized incorporating awareness of cultural context to women's vulnerability and gender sensitivity in the disaster planning of Pakistan (Bari 1998). It could be expected that these programmes might help to increase awareness of gender and reduce the gender differences and discrimination in society thus reducing the vulnerability of local women and female responders during disasters. It would also inspire the responders and female staff and increase their presence at the field level after disasters. As Fordham mentioned in her research, 'This combination of resistance and empowerment is indicative of a latent resource which can be usefully deployed to extend and improve disaster management' (Fordham 1998, p. 140).

Women and disaster knowledge

To improve disaster knowledge among the female inhabitants, schools could arrange awareness programmes, swimming lessons and walking tours to the cyclone shelters and other safe shelters. All these programmes should be arranged especially at the primary school level, where girls' attendance is higher. Again, it should be remembered that awareness programmes only for women might not bring much change for them in a patriarchal society. Awareness programmes for young girls and women should also be arranged at the house level, including their parents, husbands or other senior male members, as Connell (2003) suggested, to involve 'men and boys in the promotion of gender equality' (Connell 2003, p. 2). Besides, including male members from the family will be helpful to increase the number of the female participants in these awareness programmes.

Women's decision-making power

Patriarchal society has a strong influence on women's decision-making power, even in choosing their own dress or getting treatment for their health problems. The research has revealed that, despite its severe drawbacks, women in such a traditional society are bound to follow these social rules. For example, Bangladeshi married women in remote rural areas have to wear a sari, despite its several life-threatening problems in disasters. When arranging awareness programmes, this topic could be included in discussions on women's decision-making power.

Awareness and family disaster management plans

People frequently face mismanagement within the family in relation to disaster preparations. Supporting them to make a family work plan on responsibilities during disasters would reduce vulnerability and coordination difficulties and increase efficient evacuation. Local NGOs could help in making this plan. For instance, if a family has a fixed plan in choosing shelters (houses of relatives or friends or neighbours), travel plans and having extra food or support for pregnant women, children, older and disabled people, it will help them avoid confusion and dilemmas and improve their effort to save all the members of the family.

Working together: government, NGOs and locals

Coordination difficulties and mismanagement in response programmes have been noticed in several disasters in the world. This was a problem for Bangladesh during previous cyclones, as well as other areas, like the USA after Hurricane Katrina (Branscomb and Michel-Kerjan 2006, p 396). Responders need to work in 'creative partnerships' to solve this problem and reduce the impacts of natural disasters (Begum 1993; Kunreuther et al. 2009, p. 22). Working together needs professionalism, devotedness, honesty, efficiency and enthusiasm, and Governments should participate in building trust among new/foreign responders during the emergency periods (Branscomb and Michel-Kerjan 2006; DDM 2013) and create a good working environment for them. Again, governments should increase monitoring at the field level and create more opportunities for NGOs to be prepared with sufficient information and manpower to help them for successful response programmes.

Local community and coping capacity

Integrating coping capacity of the local community will solve many insufficiencies of the disaster management plans and difficulties of implementation. Parvin mentioned in her research that 'it is necessary to build a bridge between the efforts taken at the community level and development organizations' (Parvin et al. 2008, p. 192). Some suggestions can be made in this regard.

First of all, to save money and time in building cyclone shelters, local house owners could play an important role. Traditionally, better-off people in Barguna have been the first responders to the poor victims with shelter, food and clothes during the previous cyclones. Such social relationships and assistance are also revealed among the communities in countries such as Fiji, where traditional kin leaders share food aid with others and elite capture is not found (Takasaki 2011, p. 411). In the traditional society of Barguna, if wealthier householders could be listed as social workers and introduced as such to the local society by the government, then poor, vulnerable people could count on them for shelter and support. At the same time, these local people will feel honoured, inspired and 'gain a sense of commitment' (Adnan 1992; Mallick 2014, p. 668). This plan will save a large number of people, especially women, children and disabled people, from the health impacts of disasters and decrease pressure on the limited number of cyclone shelters.

Local inhabitants and power relations

Distributing relief among the most affected poor victims gets top priority during the post-disaster phase. However, disrupted transport, illegal pressure from local influential people and political pressure from the Government and opposition parties create constraints on the distribution of relief. This chaos ends up in the exclusion of the genuinely poor victims from the list. To improve the situation, empowering the general population, who are well aware of the power relations among local influential people, and recruiting them to the Government disaster management committee would play a significant role in decision-making for disaster management to be effective (Murtaza et al. 2013; Dominelli 2014).

Constructing more infrastructure

Preparing for disasters before the emergency creates helpful environments for implementing disaster management plans. The construction of river embankments, the provision of a sufficient number of shelters and well-prepared healthcare facilities decrease the vulnerability of the disaster-prone areas and help management plans to work efficiently. The Bangladesh government, NGOs, INGOs and local wealthy residents should take the initiative in this regard. Some recommendations in this regards are rehearsed in the following paragraphs.

More cyclone shelters and embankments

The disaster-prone coastal region of Bangladesh needs more cyclone shelters and embankments. Most areas do not have a sufficient number of cyclone shelters according to their needs. Many cyclone shelters also do not have 'killa' (space for livestock). Beside this, the location of the cyclone shelters and connecting road networks become unusable during cyclones. So, more attention from the Government is needed for increasing the number and proper planning for the location of cyclone shelters. Similar suggestions for cyclone shelters were also mentioned in other research, such as in Paul and Dutt (Paul and Dutt 2010).

Many highly vulnerable areas are not protected with embankments. These areas will be easily affected by future cyclones. The location of the residential areas is often so close to the rivers that no other preparation plans will be successful to save all of the inhabitants unless there is some protection from tidal surges. So, constructing embankments is highly recommended to save these inhabitants from cyclone disasters.

Improving the transport system

Transport systems are very poor in rural areas. Many disaster-prone villages have only waterways and local earthen roads which become inaccessible after disasters. Government should consider these remote areas as a matter of importance, and take proper steps to improve the transport systems. Roads and proper launch terminals should be built in those areas.

Improvement of prevailing healthcare facilities

Lack of local healthcare facilities becomes one of the major problems in healthcare access during disasters. Developing countries face this problem most intensely, especially in remote areas like Barguna. This area and other coastal regions of Bangladesh face a lack of medical staff, essential equipment and other facilities, along with twin issues of mismanagement and corruption. To improve these conditions, more local healthcare centres need to be established in the remote areas, and Government and private organizations should take the proper steps to improve the prevailing healthcare centres. Management committees should be stronger to stop mismanagement and corruption in these centres (Adams et al. 2013).

Psychological support for victims

The psychological health impacts of disasters do not receive proper attention in developing countries in the response and recovery programmes of government and other organizations (Nahar et al. 2014). Most of the time, psychological health impacts are not recognized properly due to a lack of awareness and the unavailability of specialist medical staff. However, the experience of this thesis research shows the presence of mental health problems among different social groups. Appropriate medical interventions and psychological support are highly recommended to reduce the health impact of disasters (Markenson and Reynolds 2006; Tunstall et al. 2006; Paxson et al. 2012; Nahar et al. 2014). In this regard, mental health centres are needed in the remote, disaster-prone areas. There is a severe lack at present in Barguna. Actually, Bangladesh has only one mental health hospital for the whole country (Government of Bangladesh 2007 in Nahar et al. 2014), and the limited private sector is mostly concentrated in the mega cities like Dhaka and Chittagong. So, supporting the mental health problems of victims requires the proper attention by the Bangladesh government and private organizations to building more mental healthcare centres. Besides, training should be arranged to improve awareness among the victims at the individual level. In this case, Vetter et al. (2011) suggest introducing the 'internet-based Self-Assessment' for the urban areas, and for remote, disaster-prone coastal districts like Barguna, governments and NGOs should have proper plans and policies on a larger scale. The training of volunteers for the non-medical component could be arranged as Nahar et al. (2014) proposed in their research, such as 'the adoption of a mental health and psychosocial support framework developed by WHO after the 2004 Indian Ocean tsunami' (Nahar et al. 2014, p. 8). Rajkumar also emphasized the importance of incorporating ethno-cultural beliefs and practices to mental health interventions for disaster victims (Rajkumar et al. 2008).

8.8 Conclusion

I started this research with some questions. Being a researcher and citizen of a disaster-prone country, I wanted to know more about the gendered impacts of disasters on health and healthcare access and to find a way to lessen the impacts. I have reviewed the prevailing literature, which shows the gaps in gender and disaster research, despite its importance. I found that the gendered health impacts of disasters have not received enough attention in disaster research and disaster management, which instead focus on numbers of injuries and casualties and some of the reasons behind them. These studies do not provide any in-depth investigation of the factors affecting vulnerability, the effects of injuries and the consequences of incomplete treatments on the lives and lifestyles of the survivors, especially women.

I have followed a qualitative methodology, which helped me to reach the field level, at the doorstep of the victims. Without a close analysis of victims' experiences, it is not always clear exactly where the problems lie nor how they might be resolved. This methodology helped me to capture the deeply rooted, rarely investigated socio-cultural factors and their influence on women's health, both in everyday life and during disasters. The research has drawn out detailed aspects of cultures and, in so doing, it demonstrates why the larger patterns exist. The thesis emphasizes gender, a complex, socially constructed idea and product of culture, defines the meaning of men and women and their gendered roles (Rose 2010) and clearly shows the strong influence of culture and gender in differentiating disaster vulnerability among social members and increasing women's vulnerability, especially for those with a poor economic background. It became clear that even education and employment cannot always assure women's empowerment and control over their own decisions in a strongly traditional society, where disaster and poverty intensify the problems. These findings identify a need for the emergence of considering gender and culture issues in disaster management policy and programming (Fisher 2010; Hyndman 2008; Canon 2002) and advocate comprehensive, gender-sensitive disaster management plans (Alam and Rahman 2014) that focus on 'gender mainstreaming' (Momsen 2010, p. 15) and 'WCD' approaches highlighting women and their higher vulnerable status, along with culture and economic issues and influences (Bhavnani et al. 2003) to improve resilience against the disaster health impacts. Gender equity (Sen 2002; Doyal 2000) and equality should be a special concern for these disaster plans, considering local traditions, culture and social attitudes (Dominelli 2013) to avoid power collisions at the family and social level. Though it will be complex, still we can hope that gender-sensitive disaster management plans will be successful to lessen disaster impacts more efficiently.

Future research: suggestions and topics

The important topic, intense methodology and unique findings of the present study offer suggestions for future research. First of all, this study suggests that qualitative methods and methodologies should be followed to understand disaster impacts. A qualitative approach offers the researcher the chance to capture the problems of disaster victims closely and helps us to understand the complex inter-relations of socio-cultural factors and their influence on people's life. Secondly, more studies should be conducted focusing on gender and its relation with culture and traditions. The strong influence of social attitudes and culture on gender relations should be considered to understand people's lives and decisions, especially women's lives. Thirdly, we need research on different vulnerable disaster prone areas of the world. As the present research teaches, geographical location, socio-cultural and economic conditions and the nature of the disasters create huge differences in disaster impacts; every disaster-prone region deserves to be researched uniquely.

Again, during my long journey through the present research, I came across several important topics which in my opinion should be researched in future. These topics would be very helpful to improve disaster management planning with a better understanding of gender relations in society. First and most important is 'gender-based violence as a humanitarian crisis'. There is evidence of violence against women in disasters but such incidents remain hidden and neglected, and statistics and documentary evidence cannot be found. In my fieldwork, a number of women were found to be suffering from different types of violence, but they are not able to share their experiences, being confined with gender identity, roles and responsibilities and wishing to avoid further humiliation and insecurity. These incidents should be investigated with an intense qualitative methodology to reveal the real conditions, the reasons and how these conditions could be improved.

Secondly, a huge research gap has been identified in relation to the mental health impacts of disasters. Disaster-prone areas in developing countries still have not received proper attention for the mental health impacts of disasters despite its presence among the victims. So, research should be conducted on 'gender and mental health impacts of disasters in developing countries' to reveal the full picture, which will be helpful to lessen the health impacts of disaster more successfully.

Again, the present research teaches us the importance of considering socio-cultural factors to study women's lives in disaster-prone areas but due to time limitations, it was not possible to focus separately on vulnerable groups such as young girls, disabled women, women-headed families and divorced women. These groups need separate research and the topic could be 'disaster, culture and women' (with a special focus on young girls, disabled women, women-headed families and divorced women). It is expected that all these topics will contribute to the broader context of gender, disaster and health studies being able to understand women's lives, their vulnerability more intensively and improve their conditions and lessen unequal disaster impacts upon them.

References

a2i. (2013). "Access to Information (a2i) Programme, Prime Minister's Office." Dhaka, Bangladesh. www.azi.pmo.gov.bd, accessed on 13/3/2015.

Adams, A. M., Ahmed, T., Arifeen, S. E., Evans, T. G., Huda, T. and Reichenbach, L. (2013). "Innovation for universal health coverage in Bangladesh: A call to action." *The Lancet* 382(9910): 2104–2111.

Adnan, S. (1992). *People's Participation, NGOs, and the Flood Action Plan: An Independent Review*. Dhaka, Bangladesh: Research & Advisory Services.

Ahmed, N. (2011). "Gender and climate change: Myth vs. reality." *End Poverty in South Asia: Promoting Dialogue on Development in South Asia*, The World Bank. http://blogs.worldbank.org/endpovertyinsouthasia/gender-and-climate-change-myth-vs-reality, accessed on 30/5/2015.

Alam, E. and Collins, A. E. (2010). "Cyclone disaster vulnerability and response experiences in coastal Bangladesh." *Disasters* 34(4): 931–954.

Alam, K. and Rahman, M. H. (2014). "Women in natural disasters: A case study from southern coastal region of Bangladesh." *International Journal of Disaster Risk Reduction* 8: 68–82.

Bari, F. (1998). *Gender, Disaster, and Empowerment: A Case Study From Pakistan*, New York, Praeger.

Begum, R. (1993). "Women in environmental disasters: The 1991 cyclone in Bangladesh." *Gender & Development* 1(1): 34–39.

Bhavnani, K., Foran, J. and Kurian, P. A. (2003). "An Introduction to Women, Culture and Development." In Bhavnani, K., Foran, J. and Kurian, P.A. (eds.) *Feminist Futures: Re-Imagining Women, Culture and Development*. London, Zed Books: 1–21.

Bishop, A. (2002). *Becoming an Ally: Breaking the Cycle of Oppression in People*. London and New York, Zed books.

Braithwaite, R. L., Bianchi, C. and Taylor, S. E. (1994). "Ethnographic approach to community organization and health empowerment." *Health Education Quarterly* 21(3): 407–416.

Branscomb, L. and Michel-Kerjan, E. (2006). "Public-Private Collaboration on a National and International Scale." In Auerswald, P. E, Branscomb, L. M., Porte, T. M. L. and Michel-Kerjan, E. O. (eds.) *Seeds of Disaster, Roots of Response: How Private Action Can Reduce Public Vulnerability*. New York, Cambridge University Press: 395–403.

Butler, J. (1990). *Gender Trouble: Feminism and the Subversion of Identity*. New York, Routledge.

Cannon, T. (2002). "Gender and climate hazards in Bangladesh." *Gender & Development* 10(2): 45–50.

Cash, R. A., Halder, S. R., Husain, M., Islam, M. S., Mallick, F. H., May, M. A., Rahman, M. and Rahman, M. A. (2014). "Reducing the health effect of natural hazards in Bangladesh." *The Lancet* 382(9910): 2094–2103.

CCC. (2009). *Climate Change, Gender and Vulnerable Groups in Bangladesh*. Dhaka, Climate Chamge Cell, DoE, MoEF, Component 4b, CDMP, MoFDM: 1–82.

Connell, R. (1987). *Gender and Power*. Cambridge, Polity Press.

Connell, R. W. (2003). *The Role of Men and Boys in Achieving Gender Equality*, Brazil, United Nations, Division for the Advancement of Women.

DDM. (2013). *Disaster Preparedness Response and Recovery*. Bangladesh, Department of Disaster Management.

Dominelli, L. (2013). "Gendering Climate Change: Implications for Debates, Policies and Practices." In Alston, M. and Whittenbury, K. (eds.) *Research, Action and Policy: Addressing the Gendered Impacts of Climate Change*. London, Springer: 77–93.

Dominelli, L. (2014). "Internationalizing professional practices: The place of social work in the international arena." *International Social Work* 57(3): 258–267.

Doyal, L. (2000). "Gender equity in health: Debates and dilemmas." *Social Science & Medicine* 51(6): 931–939.

EM-DAT. (2015). www.emdat.be, accessed on 12/02/2015.

Enarson, E. (2002). "Gender Issues in Natural Disasters: Talking Points on Research Needs." *Crisis, Women and Other Gender Concerns*. ILO, Geneva, Working paper, 5–12.

Enarson, E., Fothergill, A. and Peek, L. (2007). "Gender and Disaster: Foundations and Directions." In Rodriguez, H., Quarantelli, E. L. and Dynes, R. (eds.) *Handbook of Disaster Research*. New York, Springer: 130–146.

Fernandes, A. and Zaman, M. H. (2012). "The role of biomedical engineering in disaster management in resource-limited settings." *Bulletin of the World Health Organization* 90(8): 631–632.

Fisher, S. (2010). "Violence against women and natural disasters: Findings from post-tsunami Sri Lanka." *Violence Against Women* 16(8): 902–918.

Fordham, M. (2003). "Gender, Disaster and Development." In Pelling, M. (ed.) *Natural Disasters and Development in a Globalizing World*, New York, Routledge, Taylor & Francis: 57.

Fordham, M. H. (1998). "Making women visible in disasters: Problematising the private domain." *Disasters* 22(2): 126–143.

Fuse, A. and Yokota, H. (2012). "Lessons learned from the Japan earthquake and tsunami, 2011." *Journal of Nippon Medical School* 79(4): 312–315.

Government. (2008). *Cyclone Sidr in Bangladesh: Damage, Loss and Needs Assessment For Disaster Recovery and Reconstructions*. Bangladesh, Government of Bangladesh.

Government of Bangladesh. (2007). "WHO-AIMS report on mental health system in Bangladesh." Dhaka, Ministry of Health and Family Welfare.

GTZ. (2005). *Linking Poverty Reduction and Disaster Risk Management*, Schmidt, A., Bloemertz, L. and Macamo, E. (eds.), GTZ, Bonn: 88.

Haque, U., Hashizume, M., Kolivras, K. N., Overgaard, H. J., Das, B. and Yamamoto, T. (2012). "Reduced death rates from cyclones in Bangladesh: What more needs to be done?" *Bulletin of the World Health Organization* 90(2): 150–156. doi:10.2471/BLT.11.088302

Hossain, H., Dodge, C. and Abel, F. (1992). *From Crisis to Development: Coping With Disasters in Bangladesh*, Dhaka, University Press.

Huq-Hussain, S., Islam, M. S. and Habiba, U. (2013). "Documentation of coping strategies of coastal people particularly the Persons with Disabilities (PWDs) living in Bagerhat District." Final Report submitted to Action on Disability and Development (ADD) International, Disaster Research Training and Management Centre (DRTMC), Dhaka University.

Hyndman, J. (2008). "Feminism, conflict and disasters in post-tsunami Sri Lanka." *Gender, Technology and Development* 12(1): 101–121.

Ikeda, K. (1995). "Gender differences in human loss and vulnerability in natural disasters: A case study from Bangladesh." *Indian Journal of Gender Studies* 2(2): 171–193.

Kandaswany, D. (2005). "Media forgets female face of tsunami, Global Media Monitoring Project (GMMP)." www.womensenews.org/article.cfm/dyn/aid/2390.

Kaufman, M. (1987). "The Construction of Masculinity and the Triad of Men's Violence." In M. Kaufman (ed.) *Beyond Patriarchy: Essays an Pleasure, Power and Change*. Toronto, Oxford University Press.

Kunreuther, H., Meyer, R. and Michel-Kerjan, E. (2009). *Overcoming Decision Biases to Reduce Losses From Natural Catastrophes, Risk Management and Decision Processes Center*, USA, The Wharton School of the University of Pennsylvania: 1–27.

MacDonald, R. (2005). "How women were affected by the tsunami: A perspective from Oxfam." *PLoS Medicine* 2(6): e178.

Mallick, B. (2014). "Cyclone shelters and their locational suitability: An empirical analysis from coastal Bangladesh." *Disasters* 38(3): 654–671.

Markenson, D. and Reynolds, S. (2006). "The pediatrician and disaster preparedness." *Pediatrics* 117(2): e340-e362.

Momsen, J. (2010). *Gender and Development*, New York, Routledge.

Murtaza, N., Shirin, M. and Alam, K. (2013). Enhancing disaster resilience in Borobogi Union, Barguna district, Bangladesh-Community Managed Reduction Program in Taltoli Upazilla, ACF.

Nahar, N., Blomstedt, Y., Wu, B., Kandarina, I., Trisnantoro, L. and Kinsman, J. (2014). "Increasing the provision of mental health care for vulnerable, disaster-affected people in Bangladesh." *BMC Public Health* 14(1): 1–9. www.biomedcentral.com/1471-2458/708, accessed on 17/12/2014.

Nasreen, M. (2004). "Disaster research: Exploring sociological approach to disaster in Bangladesh." *Bangladesh e-Journal of Sociology* 1(2): 1–8.

Neumayer, E. and Plümper, T. (2007). "The gendered nature of natural disasters: The impact of catastrophic events on the gender gap in life expectancy, 1981–2002." *Annals of the Association of American Geographers* 97(3): 551–566.

Nicholson, L. (1994). "Interpreting gender." *Signs* 20(1): 79–105.

Paolisso, M. and Leslie, J. (1995). "Meeting the changing health needs of women in developing countries." *Social Science & Medicine* 40(1): 55–65.

Parvin, G. A., Takahashi, F. and Shaw, R. (2008). "Coastal hazards and community-coping methods in Bangladesh." *Journal of Coastal Conservation* 12(4): 181–193.

Paul, B. K. and Dutt, S. (2010). "Hazard warnings and responses to evacuation orders: The case of Bangladesh's Cyclone Sidr." *Geographical Review* 100(3): 336–355.

Paxson, C., Fussell, E., Rhodes, J. and Waters, M. (2012). "Five years later: Recovery from post traumatic stress and psychological distress among low-income mothers affected by Hurricane Katrina." *Social Science & Medicine* 74(2): 150–157.

Phoenix, A. and Pattynama, P. (2006). "Intersectionality." *European Journal of Women's Studies* 13(3): 187–192.

The Prothom Alo. (2015a). "Women's development has been achieved but not the empowerment." www.prothom-alo.com, accessed on 1/3/2015.

The Prothom Alo. (2015b). "Women's health but family's decision." www.prothom-alo.com, accessed on 31/3/2015.

Rahman, M. (2013). "Climate Change, disaster and gender vulnerability: A study on two divisions of Bangladesh." *American Journal of Human Ecology* 2(2): 72–82.

Rajkumar, A. P., Premkumar, T. S. and Tharyan, P. (2008). "Coping with the Asian tsunami: Perspectives from Tamil Nadu, India on the determinants of resilience in the face of adversity." *Social Science & Medicine* 67(5): 844–853.

Ray-Bennett, N. S., Collins, A., Bhuiya, A., Edgeworth, R., Nahar, P. and Alamgir, F. (2010). "Exploring the meaning of health security for disaster resilience through people's perspectives in Bangladesh." *Health & Place* 16(3): 581–589.

Rezwana, N. (2013). "Factors affecting women's accessibility to the reproductive healthcare facilities in Bangladesh." *Oriental Geographer* 54 (1): 57–69.

Rose, S. O. (2010). *What is Gender History?* Cambridge, Polity Press.

Sen, A. (2002). "Why health equity?" *Health Economics* 11(8): 659–666.

Shields, S. A. (2008). "Gender: An intersectionality perspective." *Sex Roles* 59(5): 301–311.

Sultana, F. (2010). "Living in hazardous waterscapes: Gendered vulnerabilities and experiences of floods and disasters." *Environmental Hazards* 9(1): 43–53.

Sultana, F. (2011). "Spaces of Power, Places of Hardship." In Raju, S. (ed.) *Gendered Gepgraphies: Space and Places in South Asia.* India, Oxford University Press: 293–306.

Takasaki, Y. (2011). "Targeting cyclone relief within the village: Kinship, sharing, and capture." *Economic Development and Cultural Change* 59(2): 387–416.

Tunstall, S., Tapsell, S., Green, C., Floyd, P. and George, C. (2006). "The health effects of flooding: Social research results from England and Wales." *Journal of Water and Health* 4: 365–380.

Vetter, S., Rossegger, A., Elbert, T., Gerth, J., Urbaniok, F., Laubacher, A., Rossler, W. and Endrass, J. (2011). "Internet-based self-assessment after the tsunami: Lessons learned." *BMC Public Health* 11(1): 18. www.biomedcentral.com/1471-2458/11/18.

Weisman, C. S. (1997). "Changing definitions of women's health: Implications for health care and policy." *Maternal and Child Health Journal* 1(3): 179–189.

Index

Milton Keynes UK
Ingram Content Group UK Ltd.
UKHW040059071024
449327UK00019B/680